Solar Physics Committee

Measures of Positions and Areas of Sun Spots and Faculae

on Photographs

taken at Greenwich, Dehra Dun, and Melbourne; with the deduced

heliographic longitudes and latitudes, 1878-1881

Solar Physics Committee

Measures of Positions and Areas of Sun Spots and Faculae on Photographs
taken at Greenwich, Dehra Dun, and Melbourne; with the deduced heliographic longitudes and latitudes, 1878-1881

ISBN/EAN: 9783337399368

Printed in Europe, USA, Canada, Australia, Japan

Cover: Foto ©berggeist007 / pixelio.de

More available books at **www.hansebooks.com**

SOLAR PHYSICS COMMITTEE,

DEPARTMENT OF SCIENCE AND ART,

LONDON, S.W.

MEASURES

OF

POSITIONS AND AREAS

OF

SUN SPOTS AND FACULÆ

ON

PHOTOGRAPHS

TAKEN AT

GREENWICH, DEHRA DUN, AND MELBOURNE;

WITH THE DEDUCED

HELIOGRAPHIC LONGITUDES AND LATITUDES.

1878-1881.

LONDON:

PRINTED FOR HER MAJESTY'S STATIONERY OFFICE,

BY EYRE AND SPOTTISWOODE,

PRINTERS TO THE QUEEN'S MOST EXCELLENT MAJESTY.

1891.

NOTICE.

ONE of the objects of the Solar Physics Committee since its appointment in 1879 has been to obtain a complete daily record by photographs of the Sun's surface. A general catalogue of existing solar photographs and drawings, extending from the year 1832 to 1877, was drawn up by Professor Balfour Stewart and Dr. Warren de la Rue, and published as an Appendix to the first Report. The Committee resolved that, as from the beginning of 1878 the records were more complete and continuous, they should be grouped in a separate series.

Daily photographs of the Sun had already been commenced by order of the Government of India, under the Surveyor-General at Dehra Dun, N.W. Provinces, the photographs being transmitted to Mr. Lockyer for reduction. The continuity of the record, however, was interrupted by the sudden death of Sergeant Mains, R.E., the operator who was sent from this country after having been trained at Chatham and South Kensington. A successor was subsequently appointed, and since that time there have been very few days on which no photographs were taken if the Sun were visible. Still, in July and August in each year the record was broken on account of the rainy season.

The Indian photographs are now being taken by C. F. Guthrie, under the direction of J. Eccles, M.A., Deputy-Superintendent, Survey of India.

In order to attain their object, and still further reduce the gaps in the Greenwich and Indian series, the Committee communicated with the Directors of the Observatories in the Mauritius and Australia, and cordial co-operation was promised.

As a result of the communications, both Mr. Meldrum and Mr. Ellery announced their intention of sending the negatives which were necessary to fill the gaps left in the Greenwich-Indian series for measurement by the Solar Physics Committee.

Mr. Russell, the Government Astronomer at Sydney, also promised assistance, and eventually sent 163 photographs, 146 of which were taken in 1881 and 17 in 1882. Of these there are only five which were taken on days for which there are no other photographs. The want of definition of the photographs prevented any satisfactory reductions being made, and hence they have not been entered with the others.

The series of measures given in the present publication is the result of the reductions of the photographs received from the observatories named for the years 1878–1881. Only those photographs which fill gaps in the Greenwich series have been reduced. Of the Indian photographs which fill gaps, there are 41 for the year 1878, 23 for 1879, 152 for 1880, and 160 for 1881. The Mauritius photographs for years previous to 1882 have no cross-wires, and it was not possible to completely reduce them.

4

For the years 1878–81, 695 photographs were received from Melbourne. Of these four fill gaps in 1878, 12 in 1879, 16 in 1880, and 10 in 1881. The reduction of these is included in the tables which follow.

The Greenwich reductions (extracted from the volumes of Greenwich observations) are interpolated for the sake of completeness. The sign " * " with a particular date means that the reduction was made from an Indian photograph, while " † " refers to Melbourne photographs. Where no such description is given the photographs were taken at Greenwich.

The groups of spots are numbered in the order of their appearance, the numbers being those given in the Greenwich reductions.

When there is no number in the second column it is to be understood that there is a Faculæ unaccompanied by a Spot. The positions of Faculæ relative to the Spots with which they are associated are indicated by the letters n, s, p, f, c, denoting respectively north, south, preceding, following, concentric.

The Mean Solar Time given in the first column is Greenwich Mean Solar Time.

A record of the absence of spots is given at the end.

The measurements and reductions have been made by Messrs. James, Lawrence, Greening, Mills, Hill, Fowler, Porter, Richards, McWilliam, Baxandall, Coppen, and Gregory. Mr. James has chiefly been employed as examiner and in the preparation of the copy for the press.

J. NORMAN LOCKYER.

1892, Jan. 8.

MEASURES of POSITIONS and AREAS of SPOTS and FACULÆ upon the SUN's DISK on PHOTOGRAPHS taken in the YEAR 1878.

Mean Solar Time	No. of Group, and Letter for Spot	Distance from Centre in terms of Sun's Radius	Position Angle from Sun's Axis	HELIOGRAPHIC Longitude	Latitude	SPOTS Area of UMBRA for each Spot (and for Day)	Area of WHOLE Spot (and for Day)	FACULÆ Area for each Group (and for Day)
1878.								
64·012		0·908	256·7					78
		0·892	259·7					109
		0·914	114·2					222
Jan. 7								(409)
9·048		0·939	273·0					90
Jan. 10								(90)
15·955		0·866	104·7					80
		0·896	78·7					264
Jan. 17								(344)
15·973		0·844	100·5					452
		0·873	78·2					499
Jan. 17								(951)
22·078	266a	0·312	316·1	328·2	+ 7·6	32	50	
	266b	0·274	327·8	324·1	+ 7·9	16	57	
Jan. 23						(48)	(107)	
22·089	266a	0·312	315·5	328·2	+ 7·4	24	61	
	266b	0·273	327·5	324·0	+ 7·8	19	83	
Jan. 23						(43)	(144)	
23·587	266a	0·611	290·4	330·8	+ 7·7	27	77	
	266b	0·552	294·8	326·1	+ 8·2	0	35	
Jan. 25†						(27)	(112)	
24·583	266a	0·767	284·2	331·1	+ 7·1	0	84	
	266b	0·705	288·2	325·2	+ 8·5	0	20	
Jan. 26†						(0)	(104)	
38·045	267	0·730	265·7	152·5	+ 7·6	0	15	380 e
	268	0·476	263·1	133·9	− 9·1	0	21	
Feb. 8						(0)	(36)	(380)
43·786		0·870	82·7					282
Feb. 14*								(282)
Feb. 26	268a	0·427	281·4	258·3	− 1·7	0	4	
56·736	268b	0·910	98·5	153·2	−10·7	0	8	315 f
Feb. 27*						(0)	(8)	(315)
62·013	269a	0·181	35g·1	149·9	+ 3·1	18	43	
	269b	0·209	2·3	149·3	+ 4·8	0	13	
	269c	0·207	12·5	147·2	+ 4·4	17	72	
	270	0·846	84·8	92·6	+ 0·4	3	23	404 f
Mar. 4						(38)	(151)	(404)
62·028	269a	0·181	357·9	150·0	+ 3·1	20	54	
	269b	0·213	1·2	149·3	+ 5·0	0	5	
	269c	0·208	11·4	147·3	+ 4·5	4	54	
	270	0·842	84·7	92·9	+ 0·5	4	18	357 f
Mar. 4						(28)	(131)	(357)

Mean Solar Time	No. of Group, and Letter for Spot	Distance from Centre in terms of Sun's Radius	Position Angle from Sun's Axis	HELIOGRAPHIC Longitude	Latitude	SPOTS Area of UMBRA for each Spot (and for Day)	Area of WHOLE Spot (and for Day)	FACULÆ Area for each Group (and for Day)
1878.								
62⁴·945	269a	0·288	308·7	150·5	+ 3·3	7	23	
	269c	0·264	321·4	147·0	+ 4·7	8	38	
	270	0·709	81·8	93·1	+ 0·6	0	6	728 f
Mar. 5						(15)	(67)	(728)
62·952	269a	0·291	308·4	150·5	+ 3·3	3	33	
	270	0·710	81·4	93·0	+ 0·9	0	10	570 f
Mar. 5						(3)	(43)	(570)
63·629	269a	0·425	294·9	151·1	+ 3·6	0	18	
	270	0·588	78·5	93·5	+ 0·8	2	11	
Mar. 6*						(2)	(29)	
65·890		0·740	281·9					236
		0·663	260·5					178
Mar. 8*								(414)
66·727		0·888	279·9					215
		0·859	266·6					46
Mar. 9*								(261)
68·806	271b	0·621	70·8	24·3	+ 5·9	2	18	
Mar. 11*						(2)	(18)	
69·979	271a	0·371	48·1	28·7	+ 7·3	30	130	
	271b	0·424	56·1	84·1	+ 6·9	11	40	
Mar. 12						(41)	(170)	
70·003	271a	0·369	48·3	28·4	+ 7·3	22	104	
	271b	0·427	56·7	23·5	+ 6·8	11	61	
Mar. 12						(33)	(165)	
70·803	271a	0·258	16·9	29·7	+ 7·0	19	140	
	271b	0·310	35·9	23·5	+ 7·4	3	52	
Mar. 13*						(22)	(192)	
71·989	271a	0·330	318·1	31·1	+ 7·2	27	93	
	271c	0·289	327·1	27·4	+ 6·9	0	18	
	271b	0·281	338·4	24·3	+ 8·0	0	22	
Mar. 14						(27)	(133)	
72·786	271a	0·473	300·9	31·9	+ 7·5	26	142	
	271c	0·421	304·3	28·2	+ 7·0	14	63	
	271b	0·387	314·3	24·0	+ 8·8	11	82	
Mar. 15*						(51)	(287)	
73·657	271a	0·621	291·8	31·7	+ 7·5	17	86	
	271c	0·579	293·2	28·6	+ 7·2	9	61	
	271b	0·528	300·1	23·7	+ 9·0	5	67	
Mar. 16*						(31)	(214)	
75·734	271a	0·791	285·6	32·1	+ 7·7	16	80	
	271b	0·713	291·0	24·4	+ 9·5	0	6	
Mar. 18*						(16)	(86)	

* Indian photo.   † Melbourne photo.   The Areas of Spots and Faculæ are expressed in Millionths of the Sun's visible Hemisphere.

MEASURES of POSITIONS and AREAS of SPOTS and FACULÆ upon the SUN'S DISK on PHOTOGRAPHS taken in the YEAR 1878—*continued*.

Mean Solar Time	No. of Group and Letter for Spot	Distance from Centre in terms of Sun's Radius	Position Angle from Sun's Axis	HELIOGRAPHIC Longitude	Latitude	SPOTS Area of UMBRA for each Spot (and for Day)	Area of WHOLE Spot (and for Day)	FACULÆ Area for each Group (and for Day)
1878. 76d·636 Mar. 19*		0·945	282·3					525 (525)
86·813 Mar. 29*		0·960	283·0					575 (575)
91·775 Apr. 3*		0·960	78·7					256 (256)
93·681 Apr. 5*	271	0·730 / 0·883	83·3 / 76·3	46·0	+ 0·6	0 (0)	13 (13)	356 (356)
102·768 Apr. 14*		0·892	286·0					143 (143)
118·981 May 10		0·851	96·3					112 (112)
129·016 May 10		0·860	96·3					108 (108)
145·962 May 27	272a / 272b	0·795 / 0·885	82·4 / 79·9	69·1 / 59·7	+ 5·4 / + 8·5	13 54 (67)	60 232 (292)	810sp (810)
146·110 May 27	272a / 272b	0·773 / 0·871	82·0 / 79·7	69·2 / 59·5	+ 5·6 / + 8·5	17 48 (65)	55 214 (269)	851sp (851)
146·628 May 28*	272a / 272b	0·684 / 0·797	80·6 / 78·3	69·9 / 60·4	+ 5·7 / + 8·8	0 45 (45)	67 284 (351)	580c (580)
147·637 May 29*	272a / 272b	0·493 / 0·637	76·8 / 74·9	70·3 / 60·7	+ 5·7 / + 8·9	13 50 (63)	84 318 (402)	613c (613)
148·616 May 30*	272a / 272b	0·287 / 0·448 / 0·989	67·2 / 69·3 / 88·1	70·7 / 61·1	+ 5·7 / + 8·5	8 38 (46)	59 213 (272)	200 (200)
149·910 May 31	272a / 272b / 273	0·132 / 0·211 / 0·910	341·8 / 35·5 / 85·0	71·4 / 61·9 / 3·9	+ 6·6 / + 9·3 / + 4·3	7 61 0 (68)	44 240 20 (304)	528c (528)
149·929 May 31	272a / 272b / 273	0·136 / 0·211 / 0·914	341·6 / 35·4 / 84·8	71·2 / 61·6 / 3·0	+ 6·9 / + 9·3 / + 4·5	8 46 0 (54)	32 238 6 (176)	583c (583)

Mean Solar Time	No. of Group and Letter for Spot	Distance from Centre in terms of Sun's Radius	Position Angle from Sun's Axis	HELIOGRAPHIC Longitude	Latitude	SPOTS Area of UMBRA for each Spot (and for Day)	Area of WHOLE Spot (and for Day)	FACULÆ Area for each Group (and for Day)
1878. 150d·588 June 1†	272a / 272b / 272b / 273	0·238 / 0·184 / 0·162 / 0·834	300·3 / 342·4 / 346·8 / 83·8	72·0 / 63·3 / 62·2 / 4·0	+ 6·4 / + 9·6 / + 8·5 / + 4·9	0 34 20 0 (54)	20 108 114 12 (254)	
151·780 June 2*	272a / 272b / 273	0·478 / 0·364 / 0·638	284·7 / 296·6 / 82·4	71·9 / 63·5 / 5·1	+ 6·6 / + 9·1 / + 4·6	2 19 2 (23)	14 159 12 (185)	331f (331)
152·785 June 3*	272a / 272b / 273	0·667 / 0·570 / 0·444	280·9 / 286·7 / 78·7	72·0 / 64·3 / 5·1	+ 7·0 / + 9·2 / + 4·8	3 22 3 (28)	6 81 8 (95)	
153·... June 4	272	0·763	281·4	64·7	+ 8·6	9	52	
154·589 June 5†	272	0·851	279·1	64·9	+ 7·7	0 (0)	9 (9)	1741s (1741)
156·005 June 6	272	0·978	279·9	65·9	+ 9·7	0 (0)	27 (27)	656s (656)
159·951 June 10	274	0·747	260·4	343·8	− 6·7	0 (0)	14 (14)	
162·017 June 12		0·968	263·3					200 (200)
170·906 June 21		0·905	66·0					278 (278)
171·896 June 22		0·788 / 0·903	62·3 / 85·2					137 425 (562)
175·928 June 26	275a / 275b	0·689 / 0·653 / 0·970	287·9 / 287·7 / 81·3	126·9 / 124·2	+ 14·0 / + 13·4	0 4 (4)	20 20 (40)	86 (86)
176·103 June 26	275a / 275b	0·726 / 0·684 / 0·963	287·3 / 287·0 / 81·3	127·8 / 124·4	+ 14·2 / + 13·4	12 7 (19)	46 25 (71)	136 (136)
176·900 June 27	275a / 275b	0·843 / 0·797 / 0·910	285·1 / 285·1 / 81·5	128·6 / 123·9	+ 14·1 / + 13·6	7 24 (31)	45 102 (147)	180c (180) 140 (320)
177·101 June 27	275a / 275b	0·873 / 0·823 / 0·892	284·4 / 284·4 / 81·8	129·4 / 123·9	+ 13·8 / + 13·3	11 29 (40)	68 119 (187)	322c (322) 118 (440)

* Indian photo.   The Areas of Spots and Faculæ are expressed in Millionths of the Sun's visible Hemisphere.   † Melbourne photo.

MEASURES of POSITIONS and AREAS of SPOTS and FACULÆ upon the SUN's DISK on PHOTOGRAPHS taken in the YEAR 1878—*continued.*

Mean Solar Time.	No. of Group, and Letter for Spot.	Distance from Centre in terms of Sun's Radius.	Position Angle from Sun's Axis.	Longitude.	Latitude.	Area of UMBRA for each Spot (and for Day).	Area of WHOLE for each Spot (and for Day).	Area for each Group (and for Day).
1878.			°	°	°			
177ᵈ897	275a	0·950	283·4	130·2	+13·5	29	130	176 c
	275b	0·918	283·3	124·9	+13·3	17	93	
	275c	0·904	282·8	123·0	+12·8	15	85	
		0·787	81·7					102
		0·947	98·7					114
June 28						(61)	(308)	(392)
178·106	275a	0·964	284·0	130·3	+14·2	18	131	128 c
	275b	0·932	284·1	124·3	+14·1	18	59	
	275c	0·922	283·4	122·8	+13·4	11	56	
		0·761	80·8					57
		0·929	99·8					120
June 28						(47)	(246)	(305)
178·948	275c	0·980	283·1	123·2	+13·4	15	68	137 c
June 29						(15)	(68)	(137)
179·067	275c	0·981	282·9	122·0	+13·2	0	120	61 c
June 29						(0)	(120)	(61)
195·918		0·778	76·8					368
July 16								(368)
196·111		0·768	78·7					339
July 16								(339)
196·910		0·967	75·0					120
July 17								(120)
197·115		0·940	76·5					115
July 17								(115)
205·906	276	0·590	288·6	83·2	+15·3	3	11	
		0·864	91·7					267
July 26						(3)	(11)	(267)
206·999		0·947	67·5					54
July 27								(54)
207·021		0·948	68·0					54
July 27								(54)
216·934		0·961	87·2					89
Aug. 6								(89)
217·088		0·960	87·2					190
Aug. 6								(190)
217·890		0·893	86·7					46
Aug. 7								(46)
218·104		0·860	82·2					149
Aug. 7								(149)
216·990		0·929	226·0					101
Aug. 16								(101)

Mean Solar Time.	No. of Group, and Letter for Spot.	Distance from Centre in terms of Sun's Radius.	Position Angle from Sun's Axis.	Longitude.	Latitude.	Area of UMBRA for each Spot (and for Day).	Area of WHOLE for each Spot (and for Day).	Area for each Group (and for Day).
1878.			°	°	°			
244ᵈ027	277	0·978	88·2	186·0	+ 3·3	23	146	147 nf
Sept. 2						(23)	(146)	(147)
244·077	277	0·973	87·6	186·6	+ 4·0	24	136	115 nf
Sept. 2						(24)	(136)	(115)
244·949	277	0·910	89·8	186·4	+ 3·3	34	127	298 nf
Sept. 3						(34)	(127)	(298)
245·115	277	0·887	89·4	187·2	+ 3·9	26	141	283 nf
Sept. 3						(26)	(141)	(283)
245·905	277	0·794	90·3	187·2	+ 4·2	31	165	229 f
Sept. 4						(31)	(165)	(229)
248·007	277	0·413	96·4	187·2	+ 4·0	43	145	
Sept. 6						(43)	(145)	
248·079	277	0·399	97·0	187·3	+ 3·9	35	138	
Sept. 6						(35)	(138)	
248·893	277	0·218	104·1	187·6	+ 4·1	33	148	
Sept. 7						(33)	(148)	
249·781	277	0·060	179·3	188·1	+ 3·8	35	178	
Sept. 8*						(35)	(178)	
251·060	277	0·299	258·1	188·1	+ 3·4	27	139	
Sept. 9						(27)	(139)	
252·005	277	0·498	265·9	188·4	+ 4·3	34	148	
		0·963	71·7					94
Sept. 10						(34)	(148)	(94)
252·117	277	0·525	265·5	188·7	+ 3·8	30	142	
		0·967	71·3					227
Sept. 10						(30)	(142)	(227)
252·907	277	0·671	267·4	188·8	+ 3·6	27	118	
		0·916	70·2					298
Sept. 11						(27)	(118)	(298)
253·091	277	0·707	268·8	189·3	+ 4·3	34	172	
		0·901	72·3					485
Sept. 11						(34)	(172)	(485)
253·899	277	0·824	270·1	189·2	+ 4·2	32	126	146 sf
		0·903	289·5					431
		0·819	70·3					207
Sept. 12						(32)	(126)	(784)
254·899	277	0·937	271·6	190·1	+ 4·1	23	49	407 sf
		0·919	291·5					229
Sept. 13						(23)	(49)	(636)

* Indian photo.　　The Areas of Spots and Faculæ are expressed in Millionths of the Sun's visible Hemisphere.

MEASURES of POSITIONS and AREAS of SPOTS and FACULÆ upon the Sun's Disk on PHOTOGRAPHS taken in the YEAR 1878—continued.

Mean Solar Time	No. of Group and Letter for Spot	Distance from Centre in terms of Sun's Radius	Position Angle from Sun's Axis	HELIOGRAPHIC Longitude	HELIOGRAPHIC Latitude	SPOTS Area of UMBRA for each Spot (and for Day)	SPOTS Area of WHOLE Spot (and for Day)	FACULÆ Area for each Group (and for Day)
1878.								
255ᵈ·101	277	0·951	272·0	189·9	+4·2	33	150	260 sf
		0·931	292·0			(33)	(150)	359
Sept. 13								(619)
256·014		0·981	292·3					144
		0·969	272·0					118
Sept. 14								(262)
270·695 Sept. 29*		0·983	87·6					170 (170)
271·894 Sept. 30		0·914	88·5					207 (207)
272·017 Sept. 30		0·907	89·2					302 (302)
272·933 Oct. 1		0·807	88·0					401 (401)
280·004 Oct. 8		0·901	75·5					78 (78)
280·065 Oct. 8		0·896	76·5					58 (58)
283·703 Oct. 12*		0·968	86·4					331 (331)
284·778 Oct. 13*		0·915	85·3					222 (222)
291·045		0·897	296·5					177
		0·876	106·3					134
Oct. 19								(311)
292·967 Oct. 21		0·705	269·7					227 (227)
300·805 Oct. 29*	278	0·948	80·6	163·1	+10·3	13 (13)	103 (103)	361 f (361)
301·682 Oct. 30*	278	0·861	80·3	163·4	+10·6	13 (13)	91 (91)	357 f (357)
302·704 Oct. 31*	278	0·719	79·3	163·9	+10·6	17 (17)	105 (105)	160 f (160)
303·968	278	0·504	75·9	163·4	+10·6	20	103	
		0·941	262·7					97
		0·878	129·8					98
Nov. 1						(20)	(103)	(195)
304·021	278	0·493	75·7	163·5	+10·6	17	79	
		0·942	261·7					85
		0·878	130·0					90
Nov. 1						(17)	(79)	(175)

Mean Solar Time	No. of Group and Letter for Spot	Distance from Centre in terms of Sun's Radius	Position Angle from Sun's Axis	HELIOGRAPHIC Longitude	HELIOGRAPHIC Latitude	SPOTS Area of UMBRA for each Spot (and for Day)	SPOTS Area of WHOLE Spot (and for Day)	FACULÆ Area for each Group (and for Day)
1878.								
304ᵈ·783 Nov. 2*	278	0·343	69·8	163·3	+10·6	14 (14)	79 (79)	
305·804 Nov. 3*	278	0·151	42·1	163·0	+10·4	13 (13)	71 (71)	
306·714 Nov. 4*	278	0·751	316·1	163·2	+9·9	9 (9)	65 (65)	
308·047 Nov. 5	278	0·404	286·5	161·8	+10·1	17 (17)	71 (71)	
308·819 Nov. 6*	278	0·564	283·0	163·0	+10·3	13 (13)	66 (66)	
310·008 Nov. 7	278	0·776	280·8	163·6	+10·5	20 (20)	71 (71)	
310·800	278	0·869	280·8	163·8	+10·6	16	69	224 f
		0·971	84·2			(16)	(69)	136
Nov. 8*								(360)
311·692	278	0·950	279·9	163·3	+10·4	11	37	226 f
		0·918	84·3			(11)	(37)	107
Nov. 9*								(333)
318·986 Nov. 16		0·948	62·7					166 (166)
321·983 Nov. 19		0·943	276·0					168 (168)
322·790 Nov. 20*	278a	0·982	86·1	226·1	+4·2	8 (8)	50 (50)	179 p (179)
323·828 Nov. 21*	278a	0·911	86·6	226·5	+3·9	2 (2)	19 (19)	112 p (112)
324·714 Nov. 22*	278a	0·800	85·9	226·7	+4·3	0 (0)	19 (19)	122 c (122)
331·978 Nov. 29		0·775	276·2					167 (167)
332·071 Nov. 29		0·774	275·7					174 (174)
348·803 Dec. 16*		0·959	278·0					95 (95)
349·674 Dec. 17*		0·946	291·0					65 (65)
351·939 Dec. 19*	278b	0·924	268·h	348·1	−1·7	7 (7)	36 (36)	302 f (302)

* Indian photo. The Areas of Spots and Faculæ are expressed in Millionths of the Sun's visible Hemisphere.

MEASURES of POSITIONS and AREAS of SPOTS and FACULÆ upon the SUN'S DISK on PHOTOGRAPHS taken in the YEAR 1878—*continued*.

Mean Solar Time.	No. of Group, and Letter for Spot.	Distance from Centre in terms of Sun's Radius.	Position Angle from Sun's Axis.	Heliographic Longitude.	Heliographic Latitude.	Area of UMBRA for each Spot (and for Day).	Area of WHOLE Spot for each Spot (and for Day).	Area for each Group (and for Day).
1878. 352ᵈ·668 Dec. 20*	278b	0·976	269·0	348·4	− 1·4	9 (9)	26 (26)	231 sf (231)
1879. 17·805 Jan. 19*		0·957	113·3					249 (249)
18·666 Jan. 20*		0·979 0·889	89·6 114·6					83 243 (326)
19·668 Jan. 21*		0·929 0·787 0·913	88·0 117·4 258·6					252 274 66 (592)
20·675 Jan. 22*		0·917 0·923	87·8 258·5					96 35 (131)
28·804 Jan. 30*	278c	0·960 0·850 0·678	298·2 271·8 241·1	129·9	−23·7	4 (4)	29 (29)	62 155 (217)
29·790 Jan. 31*	278c	0·937 0·818	272·0 144·9	130·5	−23·9	0 (0)	7 (7)	270 196 f (466)
30·804 Feb. 1*		0·896 0·961 0·985	248·0 95·3 66·9					152 125 110 (387)
31·721 Feb. 2*		0·960 0·888 0·948	246·6 92·2 62·5					89 84 212 (385)
32·684 Feb. 3*		0·892	59·0					385 (385)
35·668 Feb. 6*		0·971 0·718	348·1 96·8					32 140 (172)
41·934 Feb. 12		0·912 0·877	198·0 229·7					186 116 (302)
41·976 Feb. 12		0·897	297·0					100 (100)
1879. 69ᵈ·812 Mar. 12*		0·944	174·2					119 (119)
71·732 Mar. 14*		0·962 0·843 0·958	274·1 104·7 152·4					63 40 133 (238)
73·710 Mar. 16*		0·944 0·959	186·6 166·0					131 83 (214)
74·662 Mar. 17*		0·887 0·949	53·1 156·9					137 63 (200)
76·647 Mar. 19*		0·943 0·939	68·8 175·4					77 51 (138)
82·848 Mar. 25*		0·852 0·949	347·9 119·9					257 181 (438)
83·790 Mar. 26*		0·888 0·964	122·8 57·4					150 221 (371)
84·654 Mar. 27*		0·816 0·915	126·6 53·5					55 249 (304)
96·023 Apr. 7		0·921	123·7					262 (262)
96·038 Apr. 7		0·917	123·8					203 (203)
96·927 Apr. 8		0·848	126·3					204 (204)
96·975 Apr. 8		0·837	126·0					171 (171)
104·094 Apr. 15	279a 279b 279c	0·449 0·469 0·530	127·8 125·6 123·7	155·0 133·3 149·0	−21·0 −20·8 −21·9	24 1 11 (36)	101 47 34 (182)	
104·587 Apr. 16†	279a 279b 279d 279c	0·374 0·394 0·420 0·431	134·4 131·6 133·4 129·5	154·3 152·6 151·7 148·9	−20·3 −20·3 −21·8 −21·6	16 0 0 16 (32)	75 74 5 73 (227)	

* Indian photo.   The Areas of Spots and Faculæ are expressed in Millionths of the Sun's visible Hemisphere.   † Melbourne photo.

MEASURES of POSITIONS and AREAS of SPOTS and FACULÆ upon the SUN's DISK on PHOTOGRAPHS taken in the YEAR 1879—*continued.*

Mean Solar Time	No. of Group and Letter for Spot	Distance from Centre in terms of Sun's Radius	Position Angle from Sun's Axis	Longitude	Latitude	Area of UMBRA for each Spot (and for Day)	Area of WHOLE for each Spot (and for Day)	Area for each Group (and for Day)
1879. 106ᵈ·095		0·911	233·5					89
		0·818	320·7					70
	279a	c·268	191·3	154·0	−20·5	28	122	
	279c	0·290	170·0	147·7	−21·8	5	22	
		0·806	58·5					184
		0·808	8·5					54
		0·957	114·5					146
Apr. 17						(33)	(144)	(543)
106·953	279a	0·358	222·1	154·2	−20·4	30	149	
	279c	0·318	304·1	147·5	−22·0	3	20	
	280	0·893	117·0	77·1	−26·4	0	9	149 n p
Apr. 18						(33)	(178)	(149)
108·012 Apr. 19	279a	0·533	239·0	154·5	−20·4	26 (26)	135 (135)	
108·068 Apr. 19	279a	0·541	239·4	154·4	−20·4	28 (28)	120 (120)	
108·562 Apr. 20†	279n	0·622	243·1	154·3	−20·3	23 (23)	130 (130)	
109·628 Apr. 21†	279a	0·780	247·3	154·3	−20·7	23 (23)	140 (140)	670 f (670)
110·558 Apr. 22†	279a	0·887	249·2	154·2	−20·7	31 (31)	178 (178)	1538 f (1538)
111·598 Apr. 23†	279a	0·969	249·5	154·3	−21·0	0 (0)	207 (207)	1248 f (1248)
127·583	281a	0·752	67·6	181·2	+14·4	0	84	
		0·833	66·6					1106
		0·934	64·7					1130
May 9†						(0)	(84)	(2236)
128·894	281a	0·547	58·2	181·0	÷14·0	13	43	
	281b	0·643	58·0	174·8	+17·4	0	3	
		0·843	61·2					246
		0·875	112·5					364
May 10						(13)	(46)	(610)
128·955	281a	0·535	57·6	181·1	+13·9	16	46	
	281b	0·629	57·4	175·1	+17·2	0	3	
		0·818	60·0					231
		0·890	113·8					276
May 10						(16)	(49)	(507)
130·916 May 12	281a	0·283	3·5	181·8	+13·5	4 (4)	20 (20)	
131·092 May 12	281a	0·284	354·6	182·0	+13·6	3 (3)	11 (11)	
135·895 May 17		0·899	288·7					426 (426)

Mean Solar Time	No. of Group and Letter for Spot	Distance from Centre in terms of Sun's Radius	Position Angle from Sun's Axis	Longitude	Latitude	Area of UMBRA for each Spot (and for Day)	Area of WHOLE for each Spot (and for Day)	Area for each Group (and for Day)
1879. 137ᵈ·944 May 19		0·933	248·3					205 (205)
138·014 May 19		0·930	248·8					203 (203)
158·998		0·830	99·2					134
		0·894	90·5					37
June 9								(171)
163·975 June 14		0·811	240·5					52 (52)
164·054 June 14		0·841	242·3					109 (109)
176·904	282a	0·700	131·4	257·2	−15·3	22	40	
	282b	0·755	129·9	253·8	−26·9	18	65	
June 27						(40)	(105)	
177·982	282a	0·565	146·5	259·7	−25·4	25	75	
	282b	0·633	141·2	253·6	−27·0	26	70	
June 28						(51)	(145)	
178·069	282a	0·554	147·9	259·8	−25·1	16	94	
	282b	0·616	142·7	254·1	−26·7	14	46	
June 28						(30)	(140)	
179·796	282a	0·492	189·5	261·0	−26·0	21	112	
	282b	c·508	175·6	253·4	−27·4	14	75	
June 30						(35)	(187)	
180·170	282a	0·513	197·9	261·0	−16·2	17	98	
	282b	0·512	183·3	252·9	−27·6	13	73	
June 30						(30)	(171)	
180·622	282a	0·546	207·4	261·2	−26·0	24	115	
	282b	0·520	194·6	253·4	−27·1	0	97	
		0·516	191·8	251·8	−27·3	12	56	
	283	0·969	617	170·0	+28·1	0	20	
July 1†						(36)	(288)	
181·927	282a	0·697	225·8	261·4	−26·4	17	101	
	282b	0·629	216·7	252·6	−27·3	11	29	
	283	0·871	59·8	169·7	+27·6	3	9	374 n f
July 2						(31)	(139)	(374)
181·949	282a	0·698	226·0	261·3	−26·3	20	105	116 c
	282b	0·627	216·8	252·3	−27·1	7	30	
	283	0·869	59·6	169·7	+27·8	0	10	435 n f
July 2						(29)	(145)	(551)
182·587	282a	0·775	231·1	261·3	−26·7	31	128	
	282b	0·698	224·6	252·2	−27·0	0	37	
	283	0·800	56·6	169·9	+27·9	8	25	1320 n f
July 3†						(39)	(190)	(1320)

The Areas of Spots and Faculæ are expressed in Millionths of the Sun's visible Hemisphere.   † Melbourne photo.

MEASURES of POSITIONS and AREAS of SPOTS and FACULÆ upon the SUN'S DISK on PHOTOGRAPHS taken in the YEAR 1879—*continued.*

Mean Solar Time.	No. of Group, and Letter for Spot.	Distance from Centre in terms of Sun's Radius.	Position Angle from Sun's Axis.	HELIOGRAPHIC Longitude.	HELIOGRAPHIC Latitude.	Area of UMBRA for each Spot (and for Day).	Area of WHOLE Spot for each Spot (and for Day).	Area for each Group (and for Day).
1879.								
183·996	282a	0·913	239·1	261·0	−26·3	20	90	591 f
	283	0·617	47·6	169·5	+27·4	0	12·	
		0·832	51·3					315
July 4						(20)	(102)	(906)
184·900	282a	0·973	241·7	261·2	−26·5	11	66	657 sf
	283	0·497	33·2	169·6	+27·2	1	9	432 f
July 5						(12)	(75)	(1089)
184·977	282a	0·975	241·8	260·8	−26·5	4	81	710 sf
	283	0·490	33·9	169·5	+27·2	1	10	658 f
July 5						(5)	(91)	(1368)
187·039		0·912	310·0					94
		0·971	85·2					86
July 7								(180)
187·260		0·911	309·7					135
		0·970	85·0					155
July 7								(290)
194·622	283a	0·512	273·3	90·5	+5·5	23	129	
July 15†						(23)	(129)	
197·891		0·955	275·7					222
July 18								(222)
209·902		0·957	243·7					286
July 30								(286)
212·928		0·978	85·0					140
Aug. 2								(140)
222·922	284	0·849	244·8	98·7	−17·3	17	56	270 sf
	285	0·748	277·3	94·0	+9·8	26	81	
Aug. 12						(43)	(137)	(270)
223·091	284	0·869	245·4	98·9	−17·5	20	58	380 sf
	285	0·774	277·2	94·1	+9·7	70	172	
Aug. 12						(90)	(230)	(380)
223·930	284	0·943	248·8	98·9	−17·4	13	29	203 sf
	285	0·881	278·0	94·2	+10·2	52	138	
Aug. 13						(65)	(167)	(203)
238·592	285a	0·883	282·3	260·9	+14·1	0	52	814 sf
	286a	0·810	59·0	145·9	+29·1	28	121	
	286b	0·853	59·1	141·0	+29·9	19	77	1052 c
Aug. 28†						(47)	(250)	(1866)
240·910	286a	0·500	43·7	144·8	+27·8	33	162	
	286b	0·604	51·1	135·5	+28·4	6	19	
		0·906	83·0					228
Aug. 30						(39)	(181)	(228)
1879.								
241·092	286a	0·477	40·9	144·7	+27·8	26	137	
	286b	0·579	49·3	135·5	+28·4	5	16	
	286c	0·579	52·0	134·6	+27·1	0	12	
		0·905	82·3					396
Aug. 30						(31)	(165)	(396)
242·929	286a	0·366	354·7	143·2	+28·5	41	162	
	286b	0·382	7·9	137·6	+29·3	0	12	
Sept. 1						(41)	(174)	
243·095	286a	0·366	349·8	143·0	+28·2	38	155	
	286b	0·374	3·2	137·5	+29·1	0	8	
Sept. 1						(38)	(163)	
243·964	286a	0·432	327·9	142·4	+28·4	35	153	
Sept. 2						(35)	(153)	
244·060	186a	0·443	325·9	142·4	+28·4	25	156	
Sept. 2						(25)	(156)	
244·896	186a	0·544	313·5	141·6	+28·4	29	167	
Sept. 3						(29)	(167)	
245·032	186a	0·563	311·7	141·6	+28·3	32	180	
Sept. 3						(32)	(180)	
245·907	286a	0·686	304·7	141·5	+28·6	34	168	
Sept. 4						(34)	(168)	
249·904		0·829	59·5					415
Sept. 8								(415)
265·938	287	0·927	61·1	128·9	+29·3	23	96	484 n
Sept. 24						(23)	(96)	(484)
267·061	287	0·824	58·7	128·6	+29·5	12	39	352 n f
		0·927	81·2					221
Sept. 25						(12)	(39)	(573)
267·090	287	0·821	58·6	128·6	+29·5	17	56	366 n f
		0·939	80·5					131
Sept. 25						(17)	(56)	(497)
268·016	287	0·711	54·3	128·4	+29·7	27	52	
Sept. 26						(27)	(52)	
268·093	287	0·699	53·8	128·5	+29·6	13	41	176 n
Sept. 26						(13)	(41)	(176)
268·991	287	0·584	46·7	127·9	+29·6	9	29	
		0·786	77·5					500
Sept. 27						(9)	(29)	(500)
269·016	237	0·587	46·4	127·4	+29·8	7	18	
Sept. 27						(7)	(18)	

The Areas of Spots and Faculæ are expressed in Millionths of the Sun's visible Hemisphere.　　　† Melbourne photo.

MEASURES of POSITIONS and AREAS of SPOTS and FACULÆ upon the SUN'S DISK on PHOTOGRAPHS taken in the YEAR 1879—*continued.*

Mean Solar Time.	No. of Group, and Letter for Spot.	Distance from Centre in terms of Sun's Radius.	Position Angle from Sun's Axis.	Heliographic Longitude.	Heliographic Latitude.	Area of UMBRA for each Spot (and for Day).	Area of WHOLE Spot (and for Day).	Faculæ Area for each Group (and for Day).
1879.								
274ᵈ·071	288a	0·637	222·6	117·5	−22·1	4	23	
	288b	0·642	218·9	116·0	−24·0	19	31	
		0·869	67·8					335
Oct. 2						(23)	(54)	(335)
274·091	288a	0·640	222·8	117·2	−22·4	3	26	
	288b	0·647	218·8	116·0	−24·3	10	31	
		0·847	51·7					79
		0·887	65·5					41
Oct. 2						(13)	(57)	(120)
274·906	288a	0·748	232·6	118·6	−22·0	13	43	
	288b	0·734	228·3	115·6	−24·0	− 8	34	
Oct. 3						(21)	(77)	
276·070		0·851	308·8					513
		0·859	237·5					191
Oct. 4								(704)
277·980		0·987	244·0					105
Oct. 6								(105)
278·910	289a	0·969	75·4	309·5	+15·7	10	34	
	289b	0·981	72·9	306·1	+17·9	10	52	325ƒ
Oct. 7						(20)	(86)	(325)
279·051	289a	0·961	75·6	309·4	+15·6	0	23	
	289b	0·974	72·9	306·3	+18·0	7	35	633ƒ
Oct. 7						(7)	(58)	(633)
279·616	289a	0·911	75·2	310·6	+16·1	0	34	
	289b	0·931	72·4	307·6	+18·6	0	30	1800ƒ
Oct. 8†						(0)	(64)	(1800)
282·097	289a	0·568	70·9	310·3	+15·8	0	6	
	289b	0·605	70·3	307·7	+16·7	2	6	
	289b	0·638	69·5	305·4	+17·7	19	68	
	289b	0·702	68·0	300·5	+19·7	22	44	
Oct. 10						(43)	(124)	
287·039	290a	0·718	59·2	235·8	+25·8	4	28	
	290b	0·767	60·9	231·0	+23·8	0	5	
	290c	0·784	62·3	229·0	+25·1	9	23	183ƒ
Oct. 15						(13)	(56)	
287·064	290a	0·713	59·7	235·8	+25·3	0	25	
	290b	0·765	61·5	230·7	+25·3	0	5	
	290c	0·781	62·8	228·9	+24·6	9	13	959e
Oct. 15						(9)	(43)	
287·998	291a	0·466	172·9	262·6	−21·8	0	8	
	291b	0·478	167·5	259·8	−22·1	0	11	
		0·912	119·0					340
Oct. 16						(0)	(19)	(340)

Mean Solar Time.	No. of Group, and Letter for Spot.	Distance from Centre in terms of Sun's Radius.	Position Angle from Sun's Axis.	Heliographic Longitude.	Heliographic Latitude.	Area of UMBRA for each Spot (and for Day).	Area of WHOLE Spot (and for Day).	Faculæ Area for each Group (and for Day).
1879.								
288ᵈ·046	291a	0·466	174·0	262·6	−21·9	0	9	
	291b	0·478	168·6	259·7	+22·2	0	10	
		0·915	119·3					280
Oct. 16						(0)	(19)	(280)
289·957		0·890	289·0					806
	291a	0·618	120·3	266·0	−23·1	45	137	
	291c	0·595	218·5	263·9	−22·6	0	38	
	291b	0·569	214·4	260·6	−22·8	52	236	
Oct. 18						(97)	(411)	(806)
290·066		0·899	288·7					930
	291a	0·627	221·4	265·6	−23·1	49	131	
	291c	0·601	219·8	263·4	−22·4	4	56	
	291b	0·575	215·6	260·1	−22·7	53	226	
Oct. 18						(106)	(413)	(930)
291·914	291a	0·856	239·0	267·2	−22·9	51	232	}244 c
	291b	0·799	235·8	260·2	−22·9	41	211	
Oct. 20						(92)	(443)	(244)
291·951	291a	0·853	239·0	266·4	−22·7	83	265	}300 c
	291b	0·797	235·8	259·5	−22·9	52	219	
Oct. 20						(135)	(484)	(300)
292·983	291a	0·954	243·8	268·7	−23·0	51	217	}108 ƒ
	291b	0·911	241·3	260·8	−23·4	20	160	
		0·962	57·2					280
Oct. 21						(71)	(377)	(388)
296·930		0·866	236·3					50
Oct. 25								(50)
297·065		0·896	237·0					111
Oct. 25								(111)
308·941		0·922	110·3					131
Nov. 6								(131)
310·020	292a	0·495	340·5	346·9	+31·2	5	20	
	292b	0·503	345·5	344·3	+32·5	9	31	
	293	0·522	80·8	307·3	+17·8	0	5	
	294a	0·811	59·7	284·6	+26·3	0	12	}183ƒ
	294b	0·833	60·1	282·1	+26·6	3	24	
	295a	0·899	116·1	275·8	−21·5	6	22	
	295b	0·944	114·7	268·3	−21·8	44	315	}959 e
	295c	0·959	117·8	266·2	−25·4	27	92	
Nov. 7						(94)	(521)	(1142)
310·046	292a	0·488	339·8	346·7	+30·6	13	22	
	292b	0·496	345·0	344·1	+32·0	9	33	
	293	0·514	61·1	307·5	+17·5	0	8	
	294a	0·805	59·9	284·8	+26·0	6	13	}268 ƒ
	294b	0·830	60·3	281·1	+26·4	10	25	
	295a	0·898	116·3	275·6	−21·7	26	57	
	295b	0·942	114·9	272·0	−22·0	45	282	}542 c
	295c	0·939	118·1	265·9	−25·6	13	102	
Nov. 7						(116)	(542)	(810)

The Areas of Spots and Faculæ are expressed in Millionths of the Sun's visible Hemisphere.   † Melbourne photo.

MEASURES of POSITIONS and AREAS of SPOTS and FACULÆ upon the SUN'S DISK on PHOTOGRAPHS taken in the YEAR 1879—*continued.*

Mean Solar Time	No. of Group and Letter for Spot	Distance from Centre in terms of Sun's Radius	Position Angle from Sun's Axis	Longitude	Latitude	Area of UMBRA for each Spot (and for Day)	Area of WHOLE for each Spot (and for Day)	Area for each Group (and for Day)
1879. 315ᵈ·071		0·969	302·2					261
	295a	0·422	193·7	275·3	-21·2	0	14	
	295b	0·435	176·4	267·5	-22·7	60	172	
	295d	0·390	174·8	257·5	-19·8	10	21	
	295c	0·522	171·3	264·5	-28·0	0	5	
	295e	0·400	181·9	260·0	-20·5	15	58	
Nov. 12						(85)	(270)	(261)
315·924	295b	0·450	199·6	267·3	-22·2	36	149	
	295d	0·424	203·7	268·3	-19·8	17	59	
Nov. 13						(53)	(208)	
316·052	295b	0·460	203·0	267·4	-22·2	37	164	
	295d	0·437	207·9	268·8	-19·8	11	64	
Nov. 13						(48)	(228)	
316·952	295b	0·911	292·5					197
	295d	0·551	221·1	267·3	-21·9	27	163	
Nov. 14						(27)	(163)	(197)
316·997		0·846	294·7					278
	295b	0·557	221·7	267·3	-22·0	25	150	
Nov. 14						(25)	(150)	(278)
318·000	295b	0·685	233·2	266·7	-22·1	19	120	
Nov. 15						(19)	(120)	
318·006	295b	0·681	233·4	266·4	-21·8	18	102	
Nov. 15						(18)	(102)	
320·924	295b	0·972	246·5	266·5	-22·2	0	27	646 sf
		0·947	71·2					50
Nov. 18						(0)	(27)	(696)
320·983	295b	0·978	246·5	267·0	-22·4	0	23	443 sf
		0·934	70·8					89
Nov. 18						(0)	(23)	(532)
321·999		0·845	69·3					162
Nov. 19								(162)
322·009		0·834	70·7					155
Nov. 19								(155)
328·037	296	0·951	103·4	27·3	-12·2	10	28	268 sf
Nov. 25						(10)	(28)	(268)
328·071	296	0·950	103·4	27·0	-12·2	0	17	281 sf
Nov. 25						(0)	(17)	(282)
331·015		0·901	291·5					66
	296	0·628	112·9	24·7	-13·3	9	21	
	297	0·847	305·7	111·5	+30·2	0	11	146 sp
	298	0·914	113·4	355·5	-20·8	9	26	187 f
Nov. 28						(18)	(58)	(399)
1879. 331ᵈ·078								21
	296	0·917	289·8					
	297	0·624	113·2	22·7	-13·4	12	28	
	298	0·853	305·7	111·4	+30·3	0	8	78 sp
Nov. 28		0·906	113·6	355·8	-20·8	0	33	197 f
						(12)	(69)	(296)
335·055	298	0·372	164·1	359·6	-20·4	20	88	
	299	0·464	146·9	350·0	-22·3	3	18	
Dec. 2						(23)	(106)	
336·037								105
	298	0·908	259·8	0·3	-20·4	10	55	
	299	0·390	172·8	349·8	-22·3	0	10	
Dec. 3						(10)	(65)	(105)
336·963	298	0·482	221·9	0·7	-20·7	4	16	
Dec. 4						(4)	(16)	
336·970	298	0·482	221·1	0·3	-21·0	0	7	
Dec. 4						(0)	(7)	
338·927		0·700	236·8					616
		0·834	120·0					720
Dec. 6								(1336)
339·024		0·707	238·7					311
		0·811	122·5					447
Dec. 6								(758)
349·016		0·958	245·2					352
	300	0·777	113·1	131·8	-18·9	9	27	160 sf
		0·708	58·7					1961
Dec. 16		0·981	60·7					81
						(9)	(27)	(2554)
349·889	300	0·651	118·5	134·1	-19·2	23	122	
	300	0·692	118·5	130·0	-20·3	8	39	
Dec. 17*		0·926	58·6					424
						(31)	(161)	(424)
350·759	300	0·509	128·6	134·0	-19·8	22	103	
	300	0·555	125·6	130·3	-20·1	7	49	
Dec. 18*		0·883	54·5					236
						(29)	(152)	(236)
351·907	300	0·354	151·8	133·7	-19·6	24	93	
	300	0·384	143·5	129·9	-19·4	8	52	
Dec. 19*		0·747	46·2					266
						(32)	(145)	(266)
352·671	300	0·311	180·6	133·9	-19·7	19	108	
	300	0·308	167·6	129·7	-19·1	0	47	
Dec. 20*		0·942	125·4					165
						(19)	(155)	(165)

* Indian photo.　　The Areas of Spots and Faculæ are expressed in Millionths of the Sun's visible Hemisphere.

MEASURES of POSITIONS and AREAS of SPOTS and FACULÆ upon the SUN's DISK on PHOTOGRAPHS taken in the YEAR 1879—*continued.*

Mean Solar Time	No. of Group, and Letter for Spot	Distance from Centre in terms of Sun's Radius	Position Angle from Sun's Axis	Heliographic Longitude	Heliographic Latitude	Area of UMBRA for each Spot (and for Day)	Area of WHOLE Spot for each Group (and for Day)	Faculæ, Area for each Group (and for Day)
1879.			°	°	°			
353ᵈ·691	3co	0·378	216·6	134·1	−19·5	26	131	
	3oo	0·334	108·2	129·9	−18·9	0	29	109
Dec. 21*		0·879	130·0			(26)	(160)	(109)
188o.								
1·964		0·852	244·2					178
	3o1	0·930	66·6	240·2	+20·2	38	111	571 f
Jan. 3						(38)	(111)	(749)
1·988		0·861	243·2					137
	3o1	0·931	66·7	239·7	+20·2	28	103	351 f
Jan. 3						(28)	(103)	(488)
2·708		0·922	245·8					326
	3o1	0·852	64·1	241·3	−19·8	26	110	717 f
	3o1a	0·928	108·5	227·9	−18·4	0	34	
	3o1a	0·945	109·2	225·0	−19·2	9	29	356 e
Jan. 4*						(35)	(173)	(1399)
3·732		0·986	248·4					542
		0·798	312·1					112
	3o1b	0·573	52·1	254·0	+17·3	4	27	
	3o1	0·719	58·5	241·8	+19·2	39	175	
	3o1	0·756	59·7	238·4	+19·7	0	15	
	3o1	0·787	60·7	235·3	+20·1	16	62	401 f
	3o1a	0·810	110·2	229·1	−18·4	24	74	
	3o1a	0·859	111·1	223·9	−19·9	4	34	229 e
Jan. 5*						(87)	(387)	(1284)
4·824	3o1b	0·419	30·9	254·8	+17·4	6	31	
	3o1	0·562	46·8	242·2	+19·2	25	177	
	3o1	0·601	50·7	238·4	+19·1	0	14	
	3o1	0·663	52·0	233·9	+21·0	12	52	
	3o1a	0·647	113·2	229·4	−17·6	14	63	
Jan. 6*						(57)	(337)	
5·683		0·918	294·4					636
	3o1b	0·365	3·4	255·2	+17·4	4	46	
	3o1	0·449	30·4	242·7	+18·9	41	164	
	3o1	0·472	34·6	240·1	+19·0	0	70	
	3o1	0·534	40·7	234·8	+20·1	15	132	
	3o1a	0·482	120·0	230·7	−17·4	13	66	
Jan. 7*						(73)	(478)	(636)
7·609	3o1	0·428	333·0	243·0	+18·3	29	179	
	3o1	0·428	339·1	240·5	+19·4	17	70	
	3o1	0·433	348·7	236·4	+20·9	24	94	
	3o3	0·881	45·6	180·8	+35·4	0	78	
Jan. 9†						(70)	(421)	

Mean Solar Time	No. of Group, and Letter for Spot	Distance from Centre in terms of Sun's Radius	Position Angle from Sun's Axis	Heliographic Longitude	Heliographic Latitude	Area of UMBRA for each Spot (and for Day)	Area of WHOLE Spot for each Group (and for Day)	Faculæ, Area for each Group (and for Day)
188o.			°	°	°			
84ᵈ·785		0·942	240·8					365
	3o1	0·569	312·7	241·8	+18·9	63	352	
	3o1	0·537	322·7	236·0	+21·2	5	21	
	3o1c	0·489	225·5	238·0	−23·8	0	8	
	3o3	0·790	38·0	179·4	+35·0	17	82	
	3o3	0·810	39·3	176·8	+35·4	0	6	378 c
	3o3	0·827	41·0	174·1	+35·4	6	31	
		0·930	57·6					120
Jan. 10*		0·966	108·5			(91)	(500)	147 (1010)
9·769		0·983	241·2					113
	3o1	0·705	302·9	241·3	+19·1	63	323	674 c
	3o3	0·716	27·9	178·6	+35·2	7	47	
	3o3	0·730	30·2	176·1	+35·2	0	17	
	3o3	0·750	33·0	172·8	+35·1	5	41	196
Jan. 11*		0·916	111·4			(77)	(428)	(983)
10·950	3o1	0·854	295·4	241·5	+18·9	62	304	603 n f
	3o2	0·684	249·6	229·0	−17·1	13	34	
	3o2	0·659	248·6	226·9	−17·3	23	54	
	3o3	0·652	11·2	178·3	+35·4	11	46	270 c
	3o3	0·661	11·9	175·9	+35·4	0	8	
	3o3	0·672	18·2	172·3	+35·3	12	26	
	3o4	0·942	110·4	116·7	−20·7	0	5	230 s p
Jan. 12						(121)	(477)	(1103)
11·000	3o1	0·864	295·2	242·0	+19·1	52	310	518 n f
	3o2	0·698	250·0	229·6	−17·1	22	41	
	3o2	0·671	249·0	227·3	−17·2	16	42	
	3o3	0·656	9·7	178·7	+35·3	14	42	276 c
	3o3	0·667	16·9	176·4	+35·3	0	5	
	3o4	0·936	110·2	117·0	−20·4	0	8	
	3o4	0·949	104·4	114·6	−15·1	0	11	249 s p
Jan. 12						(122)	(481)	(1043)
11·807	3o1	0·934	292·5	241·5	+19·1	59	353	461 n f
	3o2	0·810	251·9	229·3	−17·2	18	88	
	3o2	0·773	251·3	225·7	−17·2	8	58	
	3o3	0·642	357·4	177·8	+35·2	0	52	
	3o3	0·638	3·9	172·8	+34·8	0	22	
Jan. 13*						(85)	(573)	(461)
12·785	3o1	0·988	290·7	241·4	+19·6	58	291	581 n f
	3o2	0·920	254·5	230·0	−16·1	61	317	
	3o2	0·888	253·4	225·4	−16·8	58	275	
	3o3	0·668	343·4	176·4	+35·2	5	26	
	3o3	0·653	349·2	171·5	+35·2	2	17	814 c
	3o5	0·295	237·4	177·7	−13·6	5	34	
	3o5	0·252	239·2	175·6	−11·9	3	36	
	3o5a	0·185	269·1	173·5	− 4·7	0	14	
Jan. 14*						(192)	(1010)	(1395)

* Indian photo.   The Areas of Spots and Faculæ are expressed in Millionths of the Sun's visible Hemisphere.   † Melbourne photo.

Measures of Positions and Areas of Spots and Faculæ upon the Sun's Disk on Photographs taken in the Year 1880—*continued*.

Mean Solar Time.	No. of Group and Letter for Spot.	Distance from Centre in terms of Sun's Radius.	Position Angle from Sun's Axis.	Heliographic Longitude.	Heliographic Latitude.	Area of UMBRA for each Spot (and for Day).	Area of WHOLE Spot (and for Day).	Area for each Group (and for Day).
1880.								
13d·707	302	0·982	255·6	230·6	−15·0	56	270	
	302	0·957	254·4	224·4	−16·2	30	174	982 c
	303	0·699	335·9	171·3	+35·1	0	19	233 c
	305	0·488	250·5	179·0	−13·5	17	102	
	305	0·424	249·0	174·8	−13·0	38	145	
	305	0·362	243·7	170·3	−13·6	12	50	
	305a	0·350	268·6	171·4	− 4·9	0	14	
Jan. 15*						(153)	(774)	(1215)
14·784		0·908	295·5					297
	305	0·693	255·9	180·2	−13·3	26	176	515
	305	0·619	254·6	174·4	−13·3	98	349	
	305	0·569	252·1	170·5	−14·1	2	24	
	306	0·343	196·8	142·9	−23·9	0	9	
	306	0·334	189·0	140·0	−24·0	4	22	
		0·970	122·0					632
Jan. 16*						(130)	(580)	(1444)
15·955		0·882	315·7					234
	305	0·872	258·8	182·0	−12·2	11	56	719 p
	305	0·798	257·6	174·1	−12·9	82	329	
	305	0·459	223·7	141·4	−23·9	10	41	
		0·895	120·5					316
Jan. 17						(103)	(426)	(1269)
16·040		0·885	314·2					240
	305	0·879	258·4	181·7	−12·6	13	78	513 p
	305	0·808	257·5	173·9	−13·0	71	360	
	306	0·478	225·5	141·9	−24·1	10	28	
		0·887	121·8					498
Jan. 17						(94)	(466)	(1251)
17·911	305	0·973	257·2	172·7	−13·6	24	63	337 c
	306	0·789	242·5	145·6	−24·7	24	88	
Jan. 19						(48)	(151)	(337)
17·978	305	0·975	257·5	172·3	−13·3	24	85	342 c
	306	0·799	242·7	145·8	−24·7	24	75	
Jan. 19						(48)	(160)	(342)
18·672	306	0·897	247·2	148·8	−22·7	3	33	881 n
	306	0·849	242·2	142·3	−24·5	7	49	
		0·893	109·5					119
Jan. 20*						(10)	(82)	(1005)
19·718	305	0·976	247·6	150·0	−23·0	0	59	454 f
		0·917	113·6					187
Jan. 21*						(0)	(59)	(641)
22·784	306a	0·761	299·2	75·4	+17·8	0	5	
		0·949	69·4					563
Jan. 24*						(0)	(5)	(563)
23·679		0·883	65·8					699
Jan. 25*								(699)

Mean Solar Time.	No. of Group and Letter for Spot.	Distance from Centre in terms of Sun's Radius.	Position Angle from Sun's Axis.	Heliographic Longitude.	Heliographic Latitude.	Area of UMBRA for each Spot (and for Day).	Area of WHOLE Spot (and for Day).	Area for each Group (and for Day).
1880.								
24d·923								299
	307	0·778	232·7					
	307	0·688	56·8	316·1	+17·5	0	6	513 f
	307	0·733	58·3	322·2	+18·3	0	8	143
		0·903	119·7					84
		0·957	114·5					(1039)
Jan. 26						(0)	(14)	
24·979		0·803	234·2					443
	307	0·682	56·8	325·8	+17·3	0	6	452 f
	307	0·728	58·3	321·9	+18·2	0	8	107
		0·911	116·8					90
		0·962	115·2					(1092)
Jan. 26						(0)	(14)	
26·042		0·916	240·2					259
		0·810	122·2					49
		0·883	116·3					69
Jan. 27								(377)
26·062		0·916	239·3					394
		0·804	122·2					65
		0·868	115·8					82
Jan. 27								(541)
27·672								296 p
Jan. 29*						(0)	(49)	(296)
29·136	308	0·961	67·6	238·6	+19·5	8	32	401 n p
	308	0·982	68·7	233·1	+19·5	114	415	
	309	0·967	104·3	233·0	−15·4	48	204	552 s f
Jan. 30						(170)	(651)	(933)
29·052	308	0·958	67·8	239·0	+19·2	11	41	555 n p
	308	0·979	68·8	233·7	+19·3	75	408	
	309	0·965	104·5	233·2	−15·6	27	183	475 f
Jan. 30						(113)	(632)	(1030)
29·921	308	0·898	64·9	238·0	+19·2	7	25	790 n p
	308	0·933	66·3	232·5	+19·5	97	471	
	308	0·971	66·3	225·0	+21·2	0	51	
	309	0·893	103·9	233·7	−15·1	25	199	433 s f
Jan. 31						(129)	(746)	(1223)
30·653	308	0·824	61·2	237·9	+19·4	9	41	
	308	0·862	62·8	233·4	+19·7	126	569	1359 r
	308	0·911	64·5	226·7	+20·2	14	94	
	309	0·809	104·7	233·6	−15·5	55	198	618 s f
Feb. 1*						(204)	(902)	(1987)
32·009		0·878	294·0					224
	308	0·655	49·9	237·8	+19·6	5	10	370 n
	308	0·700	51·9	233·9	+20·5	41	184	
	308	0·687	54·8	233·7	+18·3	35	126	
	308	0·802	55·9	224·1	+22·4	12	46	
	308	0·781	57·5	225·4	+20·4	7	18	
	309	0·604	107·4	233·3	−15·4	30	158	
Feb. 2						(130)	(542)	(·594)

* Indian photo.　　　The Areas of Spots and Faculæ are expressed in Millionths of the Sun's visible Hemisphere.

MEASURES of POSITIONS and AREAS of SPOTS and FACULÆ upon the Sun's Disk on PHOTOGRAPHS taken in the YEAR 1880—continued.

Mean Solar Time.	No. of Group, and Letter for Spot.	Distance from Centre in terms of Sun's Radius.	Position Angle from Sun's Axis.	Heliographic Longitude.	Heliographic Latitude.	Spots. Area of UMBRA for each Spot (and for Day).	Spots. Area of WHOLE Spot (and for Day).	Faculæ. Area for each Group (and for Day).
1880. 32d·016								226
								549 n
	308	0·881	294·3	238·0	+19·5	5	9	
	308	0·651	49·8	234·0	+20·4	34	181	
	308	0·697	52·0	234·0	+18·2	18	124	
	308	0·683	54·7	234·0	+18·2	0	39	
	308	0·794	55·8	124·8	+22·1	5	14	
	308	0·775	57·5	225·8	+20·1	29	155	
	309	0·602	108·0	233·4	-15·8			
Feb. 2						(91)	(522)	(775)
32·823	308	0·550	37·7	238·2	+19·9	0	16	
	308	0·595	41·2	234·4	+20·9	29	170	
	308	0·576	44·3	234·1	+18·7	29	138	
	308	0·602	43·9	232·8	+20·1	0	46	
	308	0·678	48·6	226·2	+21·3	0	48	
	309	0·451	111·9	233·5	-15·3	34	172	
	310	0·973	59·6	188·0	+27·6	25	257	
		0·968	102·5					184
Feb. 3*						(117)	(847)	(184)
34·001	308	0·486	19·2	233·8	+20·9	36	138	
	308	0·458	21·7	233·3	+18·8	24	101	
	308	0·485	23·2	231·9	+20·1	4	21	
	308	0·556	32·4	224·9	+21·9	3	19	
	309	0·236	132·6	233·3	-15·4	33	136	
	310	0·904	55·6	186·7	+27·4	68	296	911 sf
		0·880	101·3					263
Feb. 4						(168)	(711)	(1174)
34·969	308	0·463	354·6	233·5	+20·9	30	125	
	308	0·432	355·6	232·9	+19·0	23	72	
	308	0·443	358·9	231·4	+19·8	4	31	
	309	0·163	196·3	233·6	-15·4	37	146	
	310	0·814	50·2	186·5	+26·9	53	285	707 sf
		0·785	103·2					271
Feb. 5						(147)	(659)	(978)
34·987	308	0·462	354·2	233·5	+20·9	37	152	
	308	0·432	355·2	232·8	+19·0	25	87	
	308	0·444	358·8	231·2	+19·8	5	36	
	308	0·482	12·0	224·5	+21·7	2	12	
	309	0·164	196·1	233·3	-15·4	29	141	
	310	0·814	50·1	186·4	+26·9	70	281	1144 sf
		0·843	104·3					199
Feb. 5						(168)	(709)	(1343)
35·750	308	0·503	335·6	233·4	+20·9	31	138	
	308	0·477	335·1	232·8	+19·2	20	69	
	308	0·474	338·9	231·0	+19·9	4	30	
	309	0·270	134·3	233·7	-15·3	26	136	
	310	0·730	43·8	186·4	+26·5	74	310	996 sf
Feb. 6*						(155)	(683)	(996)

Mean Solar Time.	No. of Group, and Letter for Spot.	Distance from Centre in terms of Sun's Radius.	Position Angle from Sun's Axis.	Heliographic Longitude.	Heliographic Latitude.	Spots. Area of UMBRA for each Spot (and for Day).	Spots. Area of WHOLE Spot (and for Day).	Faculæ. Area for each Group (and for Day).
1880. 36d·742	308	0·609	318·5	233·0	+21·2	28	164	
	308	0·587	317·2	232·4	+19·5	13	53	
	309	0·458	249·3	233·7	-15·1	28	120	
	310	0·633	31·7	185·8	+26·4	70	332	
	310	0·682	32·8	182·6	+29·0	6	15	
	310	0·668	35·9	181·5	+26·9	0	10	1118 sf
Feb. 7*						(145)	(694)	(1118)
38·882	308	0·855	299·5	231·9	+20·9	28	150	916 c
	308	0·847	297·7	231·7	+19·2	4	17	
	309	0·805	255·9	233·1	-15·2	20	71	
	311	0·510	324·5	197·4	+18·2	20	147	
	311	0·471	332·5	192·5	+18·1	22	99	
	310	0·567	351·3	184·8	+27·4	46	255	
	310	0·575	357·2	181·1	+28·3	0	11	
	310	0·565	355·4	182·9	+27·5	5	20	
Feb. 9*						(145)	(770)	(916)
39·616		0·888	300·2					925
		0·870	251·8					785
	309	0·894	255·6	233·5	-15·8	0	48	
	308	0·928	295·6	232·9	+20·7	33	112	
	311	0·622	310·8	199·2	+18·1	24	119	
	311	0·596	312·4	197·0	+17·7	0	6	
	311	0·552	317·3	192·6	+17·7	22	125	
	311	0·612	336·5	185·5	+27·6	24	148	
	310	0·602	341·0	182·4	+28·1	0	6	
Feb. 10*						(103)	(564)	(1710)
40·737		0·939	249·4					627
	308	0·987	292·3	231·2	+20·7	21	90	
	311	0·770	300·5	198·1	+18·2	44	228	
	310	0·700	303·9	191·3	+17·6	15	76	
	310	0·708	321·6	183·6	+27·8	29	182	
	310	0·661	330·5	175·6	+28·7	0	12	
	310	0·633	330·4	174·4	+27·0	0	13	
	310	0·641	334·5	172·2	+28·9	5	43	
Feb. 11*						(114)	(544)	(627)
42·011	311	0·908	294·2	198·9	+18·6	20	116	406 c
	311	0·884	294·0	195·8	+17·5	2	16	
	311	0·856	295·6	192·1	+17·8	8	40	
	310	0·834	310·8	183·8	+28·2	17	50	667 c
	310	0·785	314·5	177·4	+28·1	7	13	
	310	0·763	314·6	175·6	+27·0	0	11	
	310	0·749	319·2	172·0	+28·9	11	23	
Feb. 12						(65)	(269)	(1073)
42·060	311	0·916	293·7	199·5	+18·5	20	191	336 c
	311	0·890	293·6	196·0	+17·4	4	22	
	311	0·865	295·2	192·6	+17·8	0	39	
	310	0·842	310·0	184·6	+28·1	36	105	854 c
	310	0·793	314·1	177·6	+28·2	0	14	
	310	0·772	313·8	176·1	+27·0	0	17	
	310	0·759	318·8	172·4	+29·2	5	33	
Feb. 12						(65)	(421)	(1190)

* Indian photo.   The Areas of Spots and Faculæ are expressed in Millionths of the Sun's visible Hemisphere.   † Melbourne photo.

MEASURES of POSITIONS and AREAS of SPOTS and FACULÆ upon the SUN'S DISK on PHOTOGRAPHS taken in the YEAR 1880—*continued*.

Mean Solar Time.	No. of Group and Letter for Spot.	Distance from Centre in terms of Sun's Radius.	Position Angle from Sun's Axis.	HELIOGRAPHIC Longitude.	HELIOGRAPHIC Latitude.	SPOTS Area of UMBRA for each Spot (and for Day).	SPOTS Area of WHOLE Spot (and for Day).	FACULÆ Area for each Group (and for Day).
1880. 42d·937		0·872	260·2					158
	311	0·972	290·7	198·9	+18·2	34	126	473 s f
	311	0·955	290·3	195·2	+17·0	0	40	
	310	0·911	305·5	183·2	+28·3	0	28	965 s f
	310	0·873	306·7	177·7	+27·3	0	12	
	310	0·864	307·6	176·3	+27·5	1	13	
	310	0·831	311·7	170·9	+28·8	0	19	
		0·933	116·3					289
Feb. 13						(35)	(240)	(1885)
42·955		0·874	259·3					316
	311	0·974	290·7	199·2	+18·3	39	120	299 s f
	311	0·950	290·3	195·8	+17·2	8	38	
	310	0·917	305·4	183·9	+28·6	0	39	610 s f
	310	0·878	306·5	178·2	+27·4	0	13	
	310	0·871	307·3	177·0	+27·6	0	16	
	310	0·839	311·3	171·5	+29·1	0	15	
		0·927	115·8					343
Feb. 13						(47)	(241)	(1568)
43·560		0·905	306·9					2173
		0·861	116·1					808
Feb. 14†								(3081)
44·666		0·963	305·7					866
		0·892	63·9					193
Feb. 15*								(1059)
45·575		0·811	332·1					717
		0·885	130·6					666
Feb. 16†								(1383)
48·085		0·902	246·2					84
		0·922	55·7					161
Feb. 18								(245)
50·900 Feb. 21	312	0·969	118·3	303·8	−29·1	21 (21)	66 (66)	248 s f (248)
51·673	312a	0·720	308·1	48·0	+20·7	0	8	226 s
	312	0·919	118·8	303·9	−29·2	33	183	
	312	0·962	120·7	295·4	−31·4	4	45	706 c
Feb. 22*						(37)	(236)	(932)
52·797		0·871	242·7					167
		0·834	299·2					401
	312	0·810	120·9	303·6	−29·0	37	207	
	312	0·871	121·0	296·5	−30·3	9	52	933 c
	312	0·882	122·2	295·1	−31·6	2	22	
Feb. 23*						(48)	(281)	(1501)
53·784		0·939	243·5					130
		0·901	295·1					234
	312	0·805	123·8	302·9	−29·2	33	158	
	312	0·770	123·7	295·5	−30·2	3	26	492 c
Feb. 24*						(36)	(184)	(856)

Mean Solar Time.	No. of Group and Letter for Spot.	Distance from Centre in terms of Sun's Radius.	Position Angle from Sun's Axis.	HELIOGRAPHIC Longitude.	HELIOGRAPHIC Latitude.	SPOTS Area of UMBRA for each Spot (and for Day).	SPOTS Area of WHOLE Spot (and for Day).	FACULÆ Area for each Group (and for Day).
1880. 54·956	312	0·543	135·5	301·9	−29·2	32	165	
	312	0·617	133·1	296·1	−31·0	1	17	
Feb. 25						(33)	(182)	
55·089	312	0·523	136·2	301·3	−28·6	26	167	
	312	0·600	133·7	295·7	−30·7	6	16	
Feb. 25						(32)	(183)	
55·961	312	0·420	151·5	301·3	−28·6	28	149	
	312	0·487	147·0	296·5	−30·9	2	10	
Feb. 26						(30)	(159)	
56·057	312	0·414	153·4	301·0	−28·7	28	161	
	312	0·480	148·3	296·1	−30·9	3	20	
Feb. 26						(31)	(181)	
56·908	312	0·370	177·3	300·8	−28·8	32	138	
		0·895	62·8					296
		0·898	114·8					178
Feb. 27						(32)	(138)	(474)
56·951	312	0·370	177·8	300·5	−28·8	28	145	
		0·905	64·3					264
		0·903	110·0					384
Feb. 27						(28)	(145)	(648)
57·721	312	0·396	202·0	300·9	−28·6	54	153	
		0·839	107·8					158
		0·863	61·2					411
		0·990	112·3					284
Feb. 28*						(54)	(153)	(853)
58·668	312	0·498	220·6	300·4	−28·9	51	149	188 c
		0·838	57·6					431
		0·941	111·4					978
Feb. 29*						(51)	(149)	(1397)
60·028	312	0·675	234·5	299·5	−28·7	32	138	
	312	0·675	301·8	297·2	+15·0	9	18	
	312	0·647	305·7	293·9	+16·0	8	12	
		0·841	112·5					640
Mar. 1						(49)	(168)	(640)
60·806	312	0·773	239·1	299·2	−28·2	45	167	582 s f
	312	0·790	295·8	297·9	+15·3	3	35	118 n
	312	0·734	299·9	293·3	+16·8	2	20	
		0·753	114·7					459
		0·808	39·0					273
		0·865	61·5					335
		0·870	103·1					174
Mar. 2*						(50)	(222)	(1741)
61·668	312	0·864	241·6	298·6	−28·1	35	183	519 s f
	312	0·803	290·6	298·9	+14·7	4	33	349 c
	312	0·835	294·8	293·3	+16·9	0	30	
		0·963	54·9					706
Mar. 3*						(39)	(246)	(1584)

* Indian photo.　　The Areas of Spots and Faculæ are expressed in Millionths of the Sun's visible Hemisphere.　　† Melbourne photo.

MEASURES of POSITIONS and AREAS of SPOTS and FACULÆ upon the SUN'S DISK on PHOTOGRAPHS taken in the YEAR 1880—*continued.*

Mean Solar Time.	No. of Group, and Letter for Spot.	Distance from Centre in terms of Sun's Radius.	Position Angle from Sun's Axis.	Heliographic Longitude.	Heliographic Latitude.	Spots. Area of UMBRA for each Spot (and for Day).	Spots. Area of WHOLE for each Spot (and for Day).	Faculæ. Area for each Group (and for Day).
1880.								
62ᵈ953		0·968	289·3					269
	312	0·960	242·7	297·4	−28·2	9	175	169 *sf*
		0·906	49·2					538
Mar. 4						(9)	(175)	(976)
63·010		0·973	288·2					295
	312	0·966	242·5	298·2	−28·4	16	79	121 *sf*
		0·899	48·8					216
Mar. 4						(16)	(79)	(632)
64·062		0·813	41·0					233
		0·968	50·7					171
Mar. 5								(404)
64·744		0·875	303·2					485
	312b	0·341	235·8	215·9	−17·9	21	54	
		0·766	37·2					764
		0·929	47·?					551
Mar. 6*						(21)	(54)	(1800)
65·716		0·842	303·2					815
	312b	0·537	248·3	217·4	−17·4	14	47	
		0·742	28·7					834
		0·877	43·6					431
		0·945	64·4					107
Mar. 7*						(14)	(47)	(2188)
66·779		0·900	300·0					390
		0·815	320·7					636
	312b	0·721	253·1	217·9	−17·1	11	47	297 *sp*
	314	0·478	15·0	164·4	+20·3	4	43	
	314	0·487	21·6	160·9	+20·0	7	65	
		0·806	35·3					298
		0·947	64·0					333
Mar. 8*						(22)	(155)	(1954)
67·823		0·909	303·8					496
	312b	0·862	254·4	218·2	−17·1	6	17	581 *f*
	314	0·476	342·9	166·6	+19·9	25	116	
	314	0·475	347·6	164·3	+20·4	0	27	
	314	0·466	353·9	161·1	+20·3	19	67	
		0·872	60·9					471
Mar. 9*						(40)	(227)	(1548)
69·092	313	0·880	246·6	203·3	−23·9	3	11	228 *np*
	313	0·867	248·1	201·7	−22·5	0	9	
	314	0·617	315·5	168·7	+19·7	23	93	
	314	0·555	325·3	161·0	+20·3	4	18	
		0·971	60·0					190
Mar. 10						(30)	(131)	(418)

Mean Solar Time.	No. of Group, and Letter for Spot.	Distance from Centre in terms of Sun's Radius.	Position Angle from Sun's Axis.	Heliographic Longitude.	Heliographic Latitude.	Spots. Area of UMBRA for each Spot (and for Day).	Spots. Area of WHOLE for each Spot (and for Day).	Faculæ. Area for each Group (and for Day).
1880.								
69ᵈ952	313	0·957	248·1	204·5	−23·0	0	43	406 n
	313	0·944	249·2	201·9	−22·0	0	25	
	314	0·740	305·9	169·6	+20·2	17	78	349 n,
	314	0·652	313·6	160·2	+20·4	4	21	
		0·924	57·5					362
Mar. 11						(21)	(167)	(1117)
70·083	313	0·964	247·6	204·3	−23·5	0	49	486 n
	313	0·746	248·6	200·5	−22·5	0	39	
	314	0·758	304·6	169·8	+20·1	23	86	417 n;
	314	0·670	311·9	160·4	+20·4	2	19	
		0·912	56·5					295
Mar. 11						(25)	(193)	(1198)
70·978		0·859	314·3					254
	314	0·869	298·0	171·1	+20·0	17	53	155 c
	314	0·838	298·5	167·6	+19·1	0	6	
	314	0·807	301·9	163·3	+20·4	20	54	
	314	0·783	303·7	160·5	+20·6	0	14	
		0·843	51·2					132
Mar. 12						(37)	(127)	(541)
71·056		0·864	313·5					359
	314	0·876	297·6	171·0	+19·9	26	84	352 c
	314	0·847	299·0	167·3	+19·9	0	9	
	314	0·814	301·2	163·2	+20·1	29	71	
	314	0·792	302·9	160·6	+20·4	0	36	
		0·832	51·0					193
Mar. 12						(55)	(200)	(904)
71·969		0·922	308·3					376
	314	0·955	293·6	171·9	+20·0	0	60	519 c
	314	0·935	295·2	167·9	+20·5	0	68	
	314	0·904	296·7	162·8	+20·2	21	92	
	314	0·883	297·1	160·0	+19·8	0	14	
	315	0·484	343·0	112·2	+20·4	24	64	
	315	0·464	346·0	110·3	+19·5	0	3	
	315	0·463	348·2	109·2	+19·7	0	3	
	315	0·485	350·1	108·6	+21·3	5	10	
	316	0·954	106·7	29·9	−18·0	20	75	206 *sf*
Mar. 13						(70)	(391)	(1101)
72·055		0·933	307·5					266
	314	0·966	293·3	173·2	+20·2	0	63	295 c
	314	0·941	294·8	167·9	+20·4	0	21	
	314	0·914	296·9	163·0	+21·0	0	24	
	314	0·892	296·7	160·1	+19·9	0	6	
	315	0·489	340·6	112·3	+20·3	31	61	
	315	0·469	343·6	110·4	+19·5	0	3	
	315	0·467	347·2	108·6	+19·9	0	4	
	315	0·490	348·2	108·5	−21·4	0	9	
	316	0·946	106·8	30·2	−18·2	0	51	239 *sf*
Mar. 13						(31)	(242)	(800)

* Indian photo.   The Areas of Spots and Faculæ are expressed in Millionths of the Sun's visible Hemisphere.

MEASURES of POSITIONS and AREAS of SPOTS and FACULÆ upon the SUN's DISK on PHOTOGRAPHS taken in the YEAR 1880—continued.

Mean Solar Time	No. of Group and Letter for Spot	Distance from Centre in terms of Sun's Radius	Position Angle from Sun's Axis	Heliographic Longitude	Heliographic Latitude	Area of UMBRA for each Spot (and for Day)	Area of WHOLE Spot (and for Day)	Faculæ: Area for each Group (and for Day)
1880. 72^d·664		0·959	305·9					656
		0·859	317·8					89
		0·815	327·0					116
	314	0·988	291·8	171·8	+20·1	8	83	
	314	0·978	292·7	168·5	+20·4	26	75	725 e
	314	0·954	294·4	162·4	+20·7	17	84	
	315	0·351	326·9	113·1	+20·7	36	149	
	315	0·516	329·9	110·3	+19·6	2	23	
	315	0·550	333·1	110·0	+22·4	0	29	
	315	0·509	333·4	108·4	+20·0	2	42	
	315	0·528	335·7	107·9	+21·8	3	17	
	316	0·891	106·2	30·7	−17·7	27	76	
	316	0·928	106·3	25·4	−17·8	0	11	504 f
Mar. 14*						(121)	(589)	(2090)
73·675		0·887	319·0					124
	315	0·680	311·6	113·8	+20·8	59	235	
	315	0·633	313·9	109·9	+19·8	18	51	
	316	0·759	106·9	31·7	−17·4	24	72	
Mar. 15*						(81)	(358)	(124)
73·046	315	0·840	299·3	113·9	+19·8	35	161	223 nf
	316	0·517	111·1	32·1	−17·0	13	44	
Mar. 16						(48)	(205)	(223)
73·082	315	0·850	299·7	114·3	+20·6	20	153	118 nf
	316	0·529	111·2	31·5	−17·1	15	59	
Mar. 16						(35)	(212)	(118)
75·933	315	0·927	295·4	114·3	−20·3	24	124	340 nf
	316	0·369	118·5	31·5	−16·8	15	52	
Mar. 17						(39)	(176)	(340)
75·976	315	0·927	294·9	113·9	+19·8	21	121	311 nf
	316	0·362	119·5	31·5	−16·9	15	48	
Mar. 17						(36)	(169)	(311)
77·107	315	0·991	291·9	114·5	+20·4	0	174	307 nf
	316	0·174	155·9	31·6	−16·1	6	33	
Mar. 18						(6)	(207)	(307)
77·991		0·823	310·5					184
	316	0·107	218·5	31·8	−16·2	0	13	
		0·930	107·0					85
Mar. 19						(0)	(13)	(269)
Second photograph on this day omitted.								
79·058		0·906	304·0					205
		0·811	107·8					99
Mar. 20								(304)
79·081		0·920	303·8					318
		0·815	108·5					182
		0·973	117·2					118
Mar. 20								(618)

Mean Solar Time	No. of Group and Letter for Spot	Distance from Centre in terms of Sun's Radius	Position Angle from Sun's Axis	Heliographic Longitude	Heliographic Latitude	Area of UMBRA for each Spot (and for Day)	Area of WHOLE Spot (and for Day)	Faculæ: Area for each Group (and for Day)
1880. 83^d·924		0·903	70·2					194
		0·900	110·8					204
Mar. 25								(398)
83·993		0·811	64·2					110
		0·860	112·5					152
Mar. 25								(262)
85·717	316a	0·408	194·1	288·8	−29·9	3	6	
	316a	0·422	188·4	286·3	−31·3	2	12	
		0·979	115·8					449
Mar. 27*						(5)	(18)	(449)
87·675	316a	0·638	234·1	291·8	−27·3	4	14	
	316a	0·628	232·0	290·4	−28·2	8	28	
	316a	0·607	226·2	286·8	−30·5	8	45	
		0·853	115·5					758
Mar. 29*						(20)	(87)	(758)
88·812	316a	0·787	241·4	291·9	−26·4	21	60	
	316a	0·748	232·5	285·5	−31·7	3	54	
	317	0·949	69·3	173·2	+17·3	54	275	499 c
Mar. 30*						(78)	(389)	(499)
89·803	316a	0·900	244·1	292·4	−26·1	2	13	
	316a	0·863	241·8	287·1	−27·5	12	47	775 c
	316	0·855	235·6	284·9	−32·5	2	24	
	317	0·862	65·8	173·2	−17·0	34	205	1033 nf
Mar. 31*						(50)	(289)	(1808)
91·001		0·949	240·8					624
	317	0·719	57·5	173·1	+17·8	27	138	268 p
Apr. 1						(27)	(138)	(892)
91·018		0·957	240·7					536
	317	0·948	236·8	283·9	−33·5	0	16	
	317	0·716	57·2	173·3	+17·9	35	137	285 p
Apr. 1						(35)	(153)	(821)
91·725		0·983	239·2					321
	318	0·397	3·5	200·6	+16·9	0	18	
	318	0·403	8·4	198·5	+17·1	2	15	
	317	0·615	48·6	173·1	+18·3	35	160	
Apr. 2*						(37)	(193)	(321)
92·915		0·911	297·3					99
	318	0·469	324·0	203·9	+16·2	9	24	
	318	0·462	330·6	200·9	+17·6	13	28	
	318	0·435	333·9	198·8	+16·8	0	3	
	317	0·467	28·6	173·7	+18·0	24	137	
Apr. 3						(46)	(192)	(99)

* Indian photo.　　　The Areas of Spots and Faculæ are expressed in Millionths of the Sun's visible Hemisphere.

MEASURES of POSITIONS and AREAS of SPOTS and FACULÆ upon the SUN's DISK on PHOTOGRAPHS taken in the YEAR 1880—*continued.*

Mean Solar Time.	No. of Group, and Letter for Spot.	Distance from Centre in terms of Sun's Radius.	Position Angle from Sun's Axis.	Heliographic Longitude.	Heliographic Latitude.	Spots. Area of UMBRA for each Spot (and for Day).	Spots. Area of WHOLE Spot (and for Day).	Faculæ. Area for each Group (and for Day).
1880. 93d·010		0·913	296·5					171
	318	0·481	322·4	203·7	+16·3	16	49	
	318	0·470	328·4	200·9	+17·5	16	31	
	318	0·442	331·5	198·7	+16·7	0	4	
	317	0·459	25·6	174·0	+18·2	24	121	
Apr. 3						(56)	(205)	(171)
94·952		0·946	245·8					124
		0·685	242·8					395
	318	0·767	297·5	205·4	+16·4	5	7	308 c
	318	0·730	301·6	201·9	+17·5	6	13	
	317	0·469	331·3	174·0	+18·2	32	124	
	317	0·487	335·4	172·6	+20·3	13	53	
		0·840	60·5					209
		0·894	117·8					72
Apr. 5						(56)	(197)	(1108)
95·083		0·927	245·0					150
		0·706	240·8					338
	318	0·782	296·8	205·2	+16·4	0	7	283 c
	318	0·748	300·3	201·2	+17·6	0	13	
	317	0·485	328·5	174·1	+18·4	25	105	
	317	0·504	332·3	173·1	+20·5	11	64	
		0·837	60·8					198
		0·889	118·0					58
Apr. 5						(36)	(189)	(1027)
96·916		0·950	291·0					192
		0·901	247·7					485
		0·872	212·2					99
♉.	317	0·737	302·7	175·3	+18·9	34	166	
Apr. 7						(34)	(166)	(776)
96·977		0·944	290·5					211
		0·878	245·8					362
	317	0·724	303·1	173·3	+18·6	44	192	
Apr. 7						(44)	(192)	(573)
98·000		0·951	247·5					344
	317	0·859	206·5	174·4	+19·1	0	24	393 c
	317	0·859	206·4	174·4	+19·0	24	93	
		0·974	118·3					64
Apr. 8						(24)	(117)	(801)
98·053		0·956	246·7					315
	317	0·867	295·9	174·8	+18·9	0	17	459 c
	317	0·867	296·0	174·8	+19·0	17	71	
		0·945	118·5					31
Apr. 8						(17)	(88)	(805)
98·754	317	0·929	292·6	174·7	+18·5	17	78	474 n
Apr. 9*						(17)	(78)	(474)

Mean Solar Time.	No. of Group, and Letter for Spot.	Distance from Centre in terms of Sun's Radius.	Position Angle from Sun's Axis.	Heliographic Longitude.	Heliographic Latitude.	Spots. Area of UMBRA for each Spot (and for Day).	Spots. Area of WHOLE Spot (and for Day).	Faculæ. Area for each Group (and for Day).
1880. 101d·919	319	0·574	243·3	101·6	−18·6	8	33	
	319	0·533	243·5	98·3	−18·6	0	10	
	319	0·509	243·8	96·9	−17·9	13	36	
	319	0·478	238·8	93·9	−19·4	38	195	
		0·901	123·0					58
Apr. 12						(59)	(274)	(58)
101·950	319	0·580	245·1	101·6	−18·8	20	45	
	319	0·535	243·8	98·3	−18·5	0	7	
	319	0·512	244·6	96·9	−17·7	10	36	
	319	0·479	239·4	93·8	−19·2	43	199	
		0·855	119·2					72
Apr. 12						(73)	(287)	(72)
102·962		0·834	302·0					237
	319	0·751	249·1	102·9	−19·3	6	11	
	319	0·726	249·1	100·7	−18·9	4	19	
	319	0·705	248·2	98·8	−19·2	0	6	
	319	0·689	249·4	97·7	−18·1	36	72	
	319	0·640	249·1	93·8	−17·5	4	9	
	319	0·633	246·1	92·7	−19·3	34	107	
Apr. 13						(84)	(224)	(237)
103·675	319	0·836	250·0	101·5	−19·7	4	51	
	319	0·799	250·2	97·7	−19·1	9	60	750 c
	319	0·751	250·5	93·2	−18·2	2	7	
	319	0·740	247·9	91·8	−19·9	32	92	
	319	0·694	249·1	88·1	−18·3	1	6	
		0·958	112·2					271
Apr. 14*						(48)	(216)	(1021)
104·786		0·935	295·7					57
	319	0·947	251·0	102·4	−19·8	10	54	
	319	0·919	251·7	97·5	−18·9	10	40	1125 c
	319	0·893	252·4	93·8	−18·1	0	22	
	319	0·877	250·6	91·7	−19·6	4	29	
		0·871	112·2					312
Apr. 15*						(24)	(145)	(1494)
105·946 Apr. 16	319	0·985	251·8	96·2	−18·9	0	42	325
						(0)	(42)	(325)
105·974 Apr. 16	319	0·993	251·8	99·2	−18·7	0	35	408 n
						(0)	(35)	(408)
106·910 Apr. 17		0·973	116·2					300
								(300)
106·959		0·971	69·2					
Apr. 17		0·971	117·0					367
								242
								(609)

* Indian photo.   The Areas of Spots and Faculæ are expressed in Millionths of the Sun's visible Hemisphere.

MEASURES of POSITIONS and AREAS of SPOTS and FACULÆ upon the SUN'S DISK on PHOTOGRAPHS taken in the YEAR 1890—*continued.*

Mean Solar Time.	No. of Group, and Letter for Spot.	Distance from Centre in terms of Sun's Radius.	Position Angle from Sun's Axis.	HELIOGRAPHIC Longitude.	HELIOGRAPHIC Latitude.	Area of UMBRA for each Spot (and for Day).	Area of WHOLE Spot (and for Day).	Area for each Group (and for Day).
1880.			°	°	°			
108ʰ·910		0·870	247·5					156
		0·829	60·3					152
Apr. 19								(308)
109·976		0·959	247·8					132
		0·856	242·3					66
		0·714	125·8					134
		0·853	132·3					258
		0·857	110·2					129
Apr. 20								(719)
109·991		0·958	250·2					234
		0·849	240·2					185
		0·686	125·3					212
		0·859	133·3					267
		0·853	110·5					189
Apr. 20								(1087)
110·898		0·777	138·5					198
		0·917	61·2					234
Apr. 21								(432)
111·590		0·854	66·4					683
Apr. 22†								(683)
113·046		0·780	245·3					306
		0·948	113·7					158
Apr. 23								(464)
114·784	320	0·943	63·8	192·0	+22·8	34	175	
	320	0·965	68·2	186·6	+19·6	20	106	1742 c
	320	0·980	67·8	182·7	+20·7	46	370	
	320	0·987	69·8	180·0	+19·1	0	74	
	320	0·995	70·0	176·4	+19·3	0	175	
Apr. 25*						(100)	(900)	(1742)
115·806	320	0·853	58·5	192·5	+23·7	41	213	
	320	0·878	62·2	188·4	+21·7	24	70	
	320	0·893	62·6	185·9	+20·5	4	116	
	320	0·908	63·4	183·6	+20·3	24	80	1805 c
	320	0·930	65·6	184·9	+19·2	9	52	
	320	0·952	66·5	175·7	+19·2	72	210	
Apr. 26*						(174)	(741)	(1805)
117·055		0·919	293·3					154
		0·839	302·3					137
		0·824	239·5					205
		0·751	218·3					173
	320	0·629	48·1	198·5	+21·1	0	52	2633 f
	320	0·704	32·2	191·7	+22·1	40	170	
	320	0·727	56·2	188·5	+20·5	23	112	
	320	0·740	58·2	186·7	+19·7	3	14	
	320	0·734	59·5	185·2	+19·4	2	68	
	320	0·777	60·2	182·9	+19·7	20	142	
	320	0·827	64·3	176·9	+18·3	66	253	
Apr. 27						(154)	(811)	(3302)

Mean Solar Time.	No. of Group, and Letter for Spot.	Distance from Centre in terms of Sun's Radius.	Position Angle from Sun's Axis.	HELIOGRAPHIC Longitude.	HELIOGRAPHIC Latitude.	Area of UMBRA for each Spot (and for Day).	Area of WHOLE Spot (and for Day).	Area for each Group (and for Day).
1880.			°	°	°			
117ʰ·089		0·924	294·2					167
		0·866	302·0					64
		0·824	239·5					234
		0·735	218·2					36
	320	0·621	47·5	198·8	+21·0	5	41	2390 f
	320	0·698	51·9	191·8	+22·0	33	165	
	320	0·723	56·1	188·4	+20·4	16	75	
	320	0·737	57·9	186·6	+19·8	1	26	
	320	0·750	59·1	185·1	+19·5	0	54	
	320	0·773	59·9	182·9	+19·8	13	118	
	320	0·827	64·1	176·6	+18·5	42	240	
Apr. 27						(110)	(719)	(2931)
117·800		0·893	242·4					37
		0·811	223·2					256
	320	0·483	32·8	202·5	+19·8	4	23	
	320	0·569	37·9	196·4	+22·7	5	27	
	320	0·604	43·0	192·2	+22·4	46	196	
	320	0·625	48·1	188·8	+21·0	31	148	
	320	0·637	50·2	187·3	+20·4	3	45	
	320	0·660	53·5	184·5	+19·6	10	97	
	320	0·678	53·7	183·1	+20·2	25	82	
	320	0·698	54·6	181·4	+20·4	0	28	
	320	0·712	57·3	179·3	+19·3	6	40	
	320	0·730	59·2	177·3	+18·8	68	272	1701 f
Apr. 28*						(198)	(958)	(1994)
118·937		0·960	242·8					205
		0·895	228·0					227
		0·864	248·2					75
	320	0·394	4·4	201·8	+19·2	0	5	
	322	0·437	9·9	199·0	+21·3	0	3	
	320	0·487	22·9	191·8	+22·6	29	76	
	320	0·489	30·2	188·4	+21·0	17	42	
	320	0·511	29·4	187·9	+22·4	0	16	
	320	0·499	32·2	187·1	+20·9	9	22	
	320	0·509	38·8	183·9	+19·4	0	20	
	320	0·526	39·3	182·9	+20·1	18	64	
	320	0·545	41·9	180·9	+20·1	0	4	
	320	0·553	44·2	179·5	+19·5	0	22	
	320	0·564	46·9	177·9	+18·9	12	75	
	320	0·585	48·7	176·0	+19·0	16	73	
Apr. 29						(101)	(420)	(507)
118·951		0·971	243·0					236
		0·890	228·5					180
		0·876	248·5					84
	320	0·436	9·4	199·0	+21·3	0	5	
	320	0·482	22·6	191·9	+22·3	22	91	
	320	0·486	30·2	188·3	+20·8	10	43	
	320	0·509	29·4	187·8	+22·3	0	18	
	320	0·494	32·1	187·2	+20·7	5	12	
	320	0·506	30·8	183·5	+18·9	1	13	
	320	0·524	39·5	182·7	+19·9	19	48	

* Indian photo.   The Areas of Spots and Faculæ are expressed in Millionths of the Sun's visible Hemisphere.   † Melbourne photo.

MEASURES of POSITIONS and AREAS of SPOTS and FACULÆ upon the SUN's DISK on PHOTOGRAPHS taken in the YEAR 1880—*continued.*

Mean Solar Time	No. of Group, and Letter for Spot	Distance from Centre in terms of Sun's Radius	Position Angle from Sun's Axis	HELIOGRAPHIC Longitude	HELIOGRAPHIC Latitude	Area of UMBRA for each Spot (and for Day)	Area of WHOLE for each Spot (and for Day)	FACULÆ Area for each Group (and for Day)
1880.			°	°	°			
118d·951	320	0·540	42·2	180·9	+19·7	0	4	
	320	0·550	44·2	179·5	+19·4	0	16	
	320	0·560	46·8	178·0	+18·7	15	79	
	320	0·584	48·7	175·9	+19·0	17	67	
Apr. 29						(89)	(396)	(500)
119·981		0·947	230·2					223
		0·887	299·8					227
	320	0·453	330·2	203·5	+19·2	4	6	
	320	0·452	355·6	192·0	+22·7	9	65	
	320	0·438	358·8	190·4	+21·9	2	20	
	320	0·429	3·8	188·1	+21·2	22	73	
	320	0·394	12·4	184·5	+18·6	0	10	
	320	0·424	15·2	183·1	+20·1	7	33	
	320	0·434	24·5	178·9	+19·2	18	90	
	320	0·449	30·6	175·9	+18·8	24	73	
Apr. 30						(86)	(370)	(450)
120·003		0·947	229·7					145
		0·901	299·7					153
	320	0·453	329·6	203·5	+19·0	0	7	
	320	0·451	355·4	191·7	+22·6	27	77	
	320	0·438	358·4	190·3	+21·9	0	7	
	320	0·427	3·5	187·9	+21·1	23	57	
	320	0·392	11·7	184·7	+18·5	0	5	
	320	0·421	14·6	183·1	+20·0	12	20	
	320	0·419	24·3	178·8	+19·0	17	78	
	320	0·444	30·2	176·9	+18·6	22	77	
Apr. 30						(101)	(337)	(298)
120·908		0·941	295·7					192
	320	0·549	315·0	201·7	+19·2	7	42	
	320	0·514	331·6	193·3	+23·6	0	13	
	320	0·502	333·6	191·5	+22·8	16	54	
	320	0·463	338·7	187·9	+21·6	7	50	
	320	0·420	348·1	182·8	+20·3	12	26	
	320	0·398	356·0	179·2	+19·4	18	47	
	320	0·369	3·9	176·0	+17·6	0	9	
	320	0·386	4·3	175·8	+18·6	14	44	
		0·792	45·7					118
May 1		0·928	67·8					101
						(74)	(285)	(411)
120·951	320	0·550	296·5					171
	320	0·554	314·2	201·7	+19·1	9	31	
	320	0·528	330·7	193·3	+23·6	0	8	
	320	0·506	332·7	191·5	+22·8	19	66	
	320	0·463	336·8	188·2	+21·3	11	33	
	320	0·419	346·8	182·8	+20·1	8	25	
	320	0·397	354·0	179·5	+19·3	18	41	
	320	0·368	2·3	176·1	+17·6	0	5	
	320	0·383	2·8	175·9	+18·5	23	38	
		0·896	40·8					68
		0·925	66·3					113
May 1		0·950	110·3					125
						(88)	(247)	(477)
1880.			°	°	°			
121d·793	320	0·683	302·8	203·0	+18·7	4	11	
	320	0·627	315·5	194·2	+23·1	0	18	
	320	0·608	318·1	192·0	+23·4	26	102	
	320	0·564	321·4	188·1	+22·5	0	12	
	320	0·496	325·1	183·3	+20·3	5	14	
	320	0·461	327·8	180·8	+19·2	9	23	
	320	0·430	332·8	178·4	+19·8	3	8	
	320	0·415	336·0	176·0	+18·5	13	24	
	321	0·978	61·6	91·6	+26·7	82	525	
May 2*	321	0·976	65·6	91·5	+23·7	0	24	460 n f
						(140)	(761)	(460)
123·003	320	0·821	297·5	200·4	+19·9	0	8	1923 c
	320	0·746	306·6	190·5	+23·6	27	132	
	320	0·634	307·9	181·8	+19·7	0	33	
	321	0·889	62·3	91·6	+22·4	0	29	1477 n f
	321	0·908	58·5	90·2	+26·5	123	432	
May 3						(150)	(634)	(3400)
123·724	320	0·888	294·1	199·4	+19·4	2	18	816 s p
	320	0·838	301·5	191·4	+23·7	8	40	
	320	0·739	300·8	182·5	+19·5	12	46	
	320	0·712	303·7	179·3	+20·4	0	19	
	321	0·804	58·7	92·6	+22·2	0	12	
	321	0·826	57·1	91·0	+24·2	11	42	
	321	0·836	55·4	90·6	+25·9	54	368	587 n f
May 4*						(87)	(545)	(1403)
124·810	320	0·936	296·9	191·3	+23·6	3	32	1911 c
	320	0·875	294·1	183·3	+19·0	2	20	
	321a	0·705	290·0	168·3	+11·4	0	14	
	321	0·699	50·0	90·2	+23·8	8	68	
	321	0·715	47·8	90·0	+25·8	58	336	
	321	0·732	50·5	87·5	+24·9	11	6	679 n f
May 5*						(72)	(476)	(2590)
126·050	321	0·528	34·7	90·7	+22·5	4	9	646 n f
	321	0·577	32·9	89·3	+25·7	4	420	
May 6						(79)	(429)	(646)
126·909		0·949	296·5					471
	321	0·505	16·7	88·9	+25·6	71	353	
	322	0·488	149·6	82·0	-27·9	0	9	
	322	0·520	145·6	78·7	-28·4	2	10	
May 7						(73)	(372)	(471)
126·921		0·951	297·2					447
		0·922	283·3					87
	321	0·507	16·4	89·0	+25·9	65	347	
	322	0·482	149·5	82·1	-27·6	3	11	
	322	0·518	145·7	78·8	-28·3	2	10	
May 7						(70)	(368)	(534)
127·936		0·932	299·8					210
	321	0·487	352·4	88·7	+25·6	86	491	
May 8						(86)	(491)	(210)

* Indian photo.　　The Areas of Spots and Faculæ are expressed in Millionths of the Sun's visible Hemisphere.

MEASURES of POSITIONS and AREAS of SPOTS and FACULÆ upon the SUN's DISK on PHOTOGRAPHS taken in the YEAR 1880—*continued.*

Mean Solar Time.	No. of Group, and Letter for Spot.	Distance from Centre in terms of Sun's Radius.	Position Angle from Sun's Axis.	HELIOGRAPHIC Longitude.	Latitude.	SPOTS Area of UMBRA for each Spot (and for Day).	Area of WHOLE Spot (and for Day).	FACULÆ Area for each Group (and for Day).
1880.								
128d·005	321	0·489	350·8	86·7	+25·6	74 (74)	473 (473)	
May 8								
128·717	521	0·534	336·1	88·2	+26·1	92 (92)	422 (422)	
May 9*								
129·686	321	0·630	321·0	87·6	+26·5	116	543	
	323	0·366	343·2	70·6	+17·4	0	26	
	323	0·361	350·7	64·9	+17·8	2 (118)	40 (609)	
May 10*								
130·946	321	0·783	307·6	88·3	+26·4	82	415	256 c
	323	0·538	309·4	70·6	+17·3	4 (86)	10 (425)	(256)
May 11								
131·053	321	0·789	307·3	87·6	+26·5	87	430	413 c
	323	0·550	308·2	70·2	+17·3	5 (92)	10 (440)	(413)
May 11								
131·905		0·879	246·2					241
		0·646	301·5					56
	321	0·876	302·2	87·7	+26·3	106	380	615 nf
	323	0·684	298·4	70·9	+16·8	3 (109)	11 (391)	(912)
May 12								
131·997	321	0·884	302·1	87·5	+26·5	96	385	532 nf
	323	0·699	298·0	71·0	+17·0	3 (99)	7 (392)	(532)
May 12								
132·905		0·930	246·0					134
		0·785	295·0					121
	321	0·955	298·3	88·0	+25·9	137	477	714 nf
	324	0·698	307·0	55·9	+22·7	5	21	60 s p
	324	0·669	310·4	52·5	+23·4	7 (149)	21 (519)	76 nf (1105)
May 13								
132·954		0·923	244·2					147
		0·793	294·5					84
	321	0·954	298·2	87·5	+25·9	101	399	805 nf
	324	0·700	306·6	56·0	+22·6	2	15	51 sp
	324	0·670	310·2	52·4	+23·4	3 (106)	17 (431)	108 nf (1195)
May 13								
133·922		0·886	291·0					146
	321	0·998	296·0	89·9	+25·7	0	542	500 nf
	324	0·796	303·2	52·1	+24·1	0 (0)	21 (563)	284 c (930)
May 14								
133·943		0·901	291·2					142
	321	0·996	296·3	87·9	+25·9	0	446	262 nf
	324	0·791	302·6	52·0	+24·2	0 (0)	14 (460)	302 c (506)
May 14								
134·922		0·968	289·0					136
		0·909	298·5					201
		0·965	118·3					198 (535)
May 15								

Mean Solar Time.	No. of Group, and Letter for Spot.	Distance from Centre in terms of Sun's Radius.	Position Angle from Sun's Axis.	HELIOGRAPHIC Longitude.	Latitude.	SPOTS Area of UMBRA for each Spot (and for Day).	Area of WHOLE Spot (and for Day).	FACULÆ Area for each Group (and for Day).
1880.								
134d·937		0·977	288·7					104
		0·913	296·3					268
		0·959	118·5					215 (587)
May 15								
135·782	·	0·970	294·3					149
		0·909	120·6					122
May 16*								(271)
136·969		0·814	126·2					210
		0·867	102·5					73
May 17								(283)
136·986		0·816	125·7					85
		0·855	102·5					194
May 17								(279)
138·691		0·936	63·5					253
May 19*								(253)
139·952	325	0·839	59·1	233·7	+24·4	9	28	246 c
	325	0·865	61·3	230·1	+23·5	0	23	
		0·839	143·0					267
		0·885	181·3			(9)	(51)	119 (632)
May 20								
140·012	325	0·830	58·6	234·0	+24·4	15	38	167 c
	325	0·841	61·1	231·0	+23·2	4	50	
		0·851	140·3					257
		0·906	176·5			(19)	(88)	65 (489)
May 20								
141·012	325	0·707	51·7	234·2	+24·6	8	22	171 c
	325	0·727	55·7	231·2	+22·9	2	14	
May 21						(10)	(36)	(171)
141·058	325	0·703	50·6	234·4	+25·1	5	13	150 c
	315	0·724	55·3	231·0	+23·0	0	7	
May 21						(5)	(20)	(150)
141·760		0·961	67·6					431
May 22*								(431)
143·722	325a	0·853	293·9	291·5	+19·5	2	7	
	325a	0·844	294·0	290·5	+19·3	2	7	107 c
	325a	0·841	295·8	289·7	+20·7	2	5	
	326	0·981	75·7	157·7	+13·8	21	74	297 c
		0·900	64·6			(27)	(93)	466 (870)
May 24*								
144·925		0·952	291·3					133
		0·883	238·5					107
		0·87?	?03·0					61
	326	0·867	75·1	161·7	+12·3	10	47	321 c
	326	0·876	74·2	160·2	+13·3	4	11	
May 15								

* Indian photo.   The Areas of Spots and Faculæ are expressed in Millionths of the Sun's visible Hemisphere.

MEASURES of POSITIONS and AREAS of SPOTS and FACULÆ upon the SUN's DISK on PHOTOGRAPHS taken in the YEAR 1880—*continued.*

Left half

Mean Solar Time	No. of Group and Letter for Spot	Distance from Centre in terms of Sun's Radius	Position Angle from Sun's Axis	Heliographic Longitude	Heliographic Latitude	Area of UMBRA for each Spot (and for Day)	Area of WHOLE Spot for each Spot (and for Day)	Faculæ Area for each Group (and for Day)
1880.								
144ᵈ·925	326	0·904	74·0	156·6	+13·9	77	286	
		0·773	153·3					60
		0·879	61·7					370
May 25						(91)	(344)	(1052)
145·071		0·985	290·8					324
		0·896	300·3					43
		0·858	236·2					175
	326	0·847	74·3	161·5	+12·6	4	58	290 c
	326	0·862	73·5	160·0	+13·5	0	17	
	326	0·892	73·5	156·3	+14·1	59	220	
		0·764	149·8					122
		0·809	59·7					520
May 25						(63)	(295)	(1474)
145·916		0·911	239·2					247
	326	0·724	71·7	162·2	+12·4	53	178	409 c
	326	0·764	72·3	158·7	+12·7	6	44	
	326	0·803	71·7	155·2	+13·9	93	379	
		0·797	58·0					301
May 26						(152)	(601)	(957)
145·991		0·911	237·2					160
	326	0·712	71·1	162·3	+12·6	48	154	208 c
	326	0·748	72·1	159·1	+12·5	8	48	
	326	0·793	71·1	153·3	+14·1	107	348	
		0·820	55·7					321
May 26						(163)	(550)	(689)
147·090	326	0·315	60·7	163·8	+13·7	50	130	
	326	0·528	64·5	162·2	+12·3	8	48	
	326	0·358	64·5	160·3	+13·1	10	33	
	326	0·622	65·0	155·8	+14·4	88	482	
		0·947						239
May 27						(156)	(693)	(239)
147·113	326	0·515	60·6	163·5	+13·8	38	120	
	326	0·531	64·0	161·8	+12·6	19	68	
	326	0·358	64·1	160·0	+13·2	8	48	
	326	0·621	64·7	155·6	+14·6	90	453	
		0·950	62·8					219
May 27						(155)	(689)	(219)
147·908	326	0·367	46·4	164·6	+13·8	21	76	
	326	0·377	53·0	162·6	+12·3	29	114	
	326	0·386	58·5	160·9	+10·8	0	30	
	326	0·416	55·1	160·0	+12·9	0	26	
	326	0·455	58·7	157·0	+12·9	27	149	
	326	0·487	57·4	155·5	+14·4	98	359	
		0·880						289
May 28						(175)	(754)	(289)
148·105		0·859	302·5					120
	326	0·343	40·0	164·7	+14·4	28	96	
	326	0·351	47·4	162·5	+12·9	38	150	
	326	0·356	53·7	160·9	+11·3	10	27	
	326	0·392	51·3	159·6	+13·4	5	24	

Right half

Mean Solar Time	No. of Group and Letter for Spot	Distance from Centre in terms of Sun's Radius	Position Angle from Sun's Axis	Heliographic Longitude	Heliographic Latitude	Area of UMBRA for each Spot (and for Day)	Area of WHOLE Spot for each Spot (and for Day)	Faculæ Area for each Group (and for Day)
1880.								
148ᵈ·105	326	0·424	55·7	156·8	+13·0	36	184	
	326	0·463	54·0	155·1	+15·0	67	348	
		0·869	59·3					357
May 28						(184)	(829)	(477)
148·914		0·920	221·3					217
		0·908	245·3					165
		0·906	300·5					210
	326	0·263	6·6	165·3	+14·4	19	50	
	326	0·254	16·7	162·8	+13·3	44	168	
	326	0·252	25·0	160·9	+12·4	0	35	
	326	0·237	33·2	159·6	+10·7	8	23	
	326	0·315	35·0	156·4	+14·2	69	592	
	326	0·351	40·7	153·3	+14·8	13	67	
		0·774	54·8					211
May 29						(153)	(935)	(803)
148·940		0·918	298·3					223
		0·918	245·2					101
		0·914	220·7					160
	326	0·262	4·9	165·5	+14·4	19	69	
	326	0·251	13·3	162·9	+13·3	25	174	
	326	0·259	26·5	160·6	+12·7	1	36	
	326	0·233	32·7	159·5	+10·6	13	31	
	326	0·313	34·6	156·3	+14·2	93	547	
	326	0·346	39·6	153·7	+14·7	16	80	
		0·777						251
May 29						(167)	(938)	(735)
149·780		0·970	245·9					149
		0·966	295·6					429
	326	0·314	325·2	166·2	+14·3	17	76	
	326	0·283	330·7	163·8	+13·7	20	98	
	326	0·305	333·9	163·6	+15·3	0	24	
	326	0·271	335·5	162·1	+13·7	23	93	
	326	0·250	332·7	162·3	+12·2	16	44	
	326	0·256	342·1	162·0	+13·4	4	34	
	326	0·210	337·7	162·0	+10·4	10	38	
	326	0·262	356·6	156·5	+14·5	132	649	
	326	0·268	7·1	153·6	+14·8	14	99	
	327	0·958	63·6	84·6	+25·0	16	82	325
May 30*						(252)	(1237)	(903)
150·785	326	0·485	301·2	167·6	+14·0	14	27	
	326	0·434	303·3	165·3	+13·9	14	30	
	326	0·438	302·0	164·7	+12·9	9	30	
	326	0·422	305·5	163·0	+13·7	10	28	
	326	0·409	303·5	162·8	+12·6	17	65	
	326	0·386	299·9	162·2	+10·6	0	8	
	326	0·396	307·5	161·2	+10·6	15	71	
	326	0·380	301·9	161·5	+11·1	4	19	
	326	0·355	301·1	160·3	+10·1	0	4	
	326	0·352	318·6	156·3	+14·8	166	755	
	327	0·880	61·3	84·4	+24·7	21	77	
	327	0·907	61·5	80·6	+25·4	4	13	350 r
May 31*						(274)	(1128)	(350)

* Indian photo.   The Areas of Spots and Faculæ are expressed in Millionths of the Sun's visible Hemisphere.

Measures of Positions and Areas of Spots and Faculæ upon the Sun's Disk on Photographs taken in the Year 1880—*continued.*

Mean Solar Time.	No. of Group, and Letter for Spot.	Distance from Centre in terms of Sun's Radius.	Position Angle from Sun's Axis.	Heliographic Longitude.	Heliographic Latitude.	Spots. Area of UMBRA for each Spot (and for Day).	Spots. Area of WHOLE Spot (and for Day).	Faculæ. Area for each Group (and for Day).
1880.								
131d·866								182
	.326	0·895	293·3					
	.326	0·682	291·2	168·8	+14·0	8	28	
	.326	0·644	291·5	166·1	+13·4	9	42	
	.326	0·610	291·5	163·4	+12·7	13	41	
	.326	0·593	292·9	162·0	+13·1	0	46	
	.326	0·585	290·6	161·8	+11·6	0	14	
	.326	0·521	300·7	155·3	+15·1	122	516	
	.327	0·754	57·2	84·3	+23·8	16	52	
	.327	0·789	57·5	81·0	+24·9	0	20	284·f
June 1*						(168)	(759)	(466)
153·116								535
	.326	0·904	299·2					
	.326	0·867	286·4	170·3	+14·0	7	13	505
	.326	0·818	286·3	165·0	+13·2	0	23	
	.326	0·736	289·4	156·9	+13·9	27	123	
	.326	0·723	291·9	155·4	+15·5	3q	154	
	.327	0·589	45·3	84·3	+24·2	15	32	
		0·913						102
		0·962						152
June 2						(88)	(345)	(1294)
153·723	.326	0·927	284·8	170·5	+13·6	2	16	
	.326	0·904	284·6	167·2	+13·1	0	17	
	.326	0·810	286·9	157·2	+13·7	18	86	
	.326	0·806	290·0	155·3	+15·9	31	131	396 n
	.327	0·515	36·3	84·0	+24·3	11	35	
		0·904	77·0					95
June 3*						(62)	(285)	(491)
155·066	.326	0·946	287·0	155·8	+16·0	0	44	867 c
	.326	0·928	285·0	152·9	+13·8	6	14	
	.327	0·388	349·8	90·0	+22·4	0	19	
June 4						(6)	(77)	(867)
156·789	.328	0·698	68·6	20·8	+14·9	10	36	
	.328	0·723	70·4	18·5	+14·1	2	8	
June 6*						(12)	(44)	
158·053	.328	0·468	58·5	21·9	+14·5	4	11	
	.328	0·515	62·1	18·3	+14·2	0	6	
June 7						(4)	(17)	
158·687	.328	0·356	46·5	22·4	+14·6	6	17	
	.328	0·379	48·6	20·8	+15·0	0	10	
	.328	0·404	52·6	18·5	+14·6	0	7	
	.329	0·991	111·7	316·6	-21·4	0	42	254 c
June 8*						(6)	(76)	(254)
159·701		0·888	295·7					224
	.328	0·243	6·2	22·9	+14·6	4	16	
	.328	0·279	22·4	18·1	+15·5	0	32	
	.328	0·272	26·2	17·3	+14·7	0	9	
	.328	0·288	27·6	16·5	+15·3	0	9	
	.329	0·938	113·1	316·7	-21·4	15	47	385 c
June 4*						(19)	(147)	(609)

Mean Solar Time.	No. of Group, and Letter for Spot.	Distance from Centre in terms of Sun's Radius.	Position Angle from Sun's Axis.	Heliographic Longitude.	Heliographic Latitude.	Spots. Area of UMBRA for each Spot (and for Day).	Spots. Area of WHOLE Spot (and for Day).	Faculæ. Area for each Group (and for Day).
1880.								
160d·985								352
	.328	0·920	300·0					
	.328	0·384	310·1	25·0	+15·0	0	5	
	.328	0·354	310·7	23·3	+14·0	0	1	
	.328	0·328	314·4	21·3	+14·0	7	15	
	.328	0·317	320·8	19·2	+14·9	5	7	
	.328	0·293	326·1	17·0	+14·7	9	78	
	.329	0·812	117·7	316·8	-21·6	14	28	451 nf
June 10						(35)	(134)	(803)
161·712		0·932	295·8					410
		0·884	311·3					348
	.328	0·459	300·1	21·8	+14·0	6	19	
	.328	0·402	306·8	17·1	+14·7	4	11	
	.328	0·397	308·6	16·4	+15·1	6	18	
	.329	0·727	121·6	316·1	-21·7	9	22	
	.329	0·740	121·2	314·8	-21·9	1	7	
June 11*						(26)	(77)	(758)
161·933	.328	0·647	292·3	19·7	+15·0	7	24	
	.328	0·605	294·4	16·2	+15·2	12	42	
June 12						(19)	(66)	
164·580		0·879	286·6					656
June 14†								(656)
165·700		0·964	285·2					214
	.330	0·951	65·9	234·1	+23·3	51	161	
	.331	0·981	65·7	226·8	+24·1	0	29	432 c
June 15*						(51)	(190)	(646)
166·733	.330	0·864	63·4	234·2	+23·4	29	139	
	.331	0·918	63·8	226·7	+24·5	0	16	
	.331	0·961	64·4	218·5	+25·0	0	21	258 f
June 16*						(29)	(176)	(258)
167·782		0·803	293·2					159
	.330	0·743	59·2	233·5	+23·4	26	160	
	.331	0·813	60·7	226·5	+24·3	7	23	
	.331	0·879	62·2	218·5	+25·0	9	25	
	.331	0·915	62·2	213·5	+25·9	10	44	480 f
	.332	0·980	109·1	200·2	-18·4	98	367	632 c
June 17*								
169·001		0·829	243·2					360
		0·976	248·2					137
	.330	0·576	49·4	232·9	+23·5	30	152	
	.331	0·661	53·7	225·6	+24·3	2	7	
	.331	0·736	57·5	218·4	+24·5	4	12	
	.331	0·752	58·3	216·8	+24·4	6	17	
	.331	0·798	58·7	212·3	+25·6	26	72	268 f
	.332	0·882	118·9	203·6	-24·3	9	12	1112 c
	.331	0·894	112·1	200·4	-18·8	89	345	
	.332	0·920	114·7	197·3	-21·8	0	29	
	.332	0·937	113·9	194·3	-21·6	0	34	
	.332	0·946	111·2	192·3	-19·4	0	24	
	.332	0·959	112·7	189·9	-21·2	0	43	
		0·943	66·7					237
June 18						(166)	(757)	(2114)

* Indian photo. The Areas of Spots and Faculæ are expressed in Millionths of the Sun's visible Hemisphere. † Melbourne photo.

MEASURES of POSITIONS and AREAS of SPOTS and FACULÆ upon the SUN's DISK on PHOTOGRAPHS taken in the

Mean Solar Time.	No. of Group and Letter for Spot.	Distance from Centre in terms of Sun's Radius.	Position Angle from Sun's Axis.	Heliographic Longitude.	Heliographic Latitude.	Area of UMBRA for each Spot (and for Day).	Area of WHOLE Spot (and for Day).	Area for each Group (and for Day).
1880. 170ᵈ·034	330	0·912	240·7	233·0	+23·5	64	329	473
	331	0·438	32·0	219·0	+29·8	0	11	
	331	0·631	41·3	218·0	+24·2	5	6	
	331	0·594	49·4	211·4	+25·6	14	52	
	331	0·673	52·5	211·1	+22·3	9	19	
	331	0·657	57·1	209·7	+16·7	0	9	
	331	0·646	65·9	200·7	−18·7	62	267	937 c
	332	0·773	116·3	200·4	−21·3	12	54	
	332	0·786	119·3	200·5	−23·2	40	66	
	332	0·791	121·6	197·7	−18·6	0	8	
	332	0·801	115·0	198·3	−22·4	7	14	
	332	0·808	119·3	198·2	−24·0	11	32	
	332	0·815	121·5	193·1	−21·7	10	27	
	332	0·851	117·1	191·7	−19·3	14	37	
	332	0·857	113·9	190·2	−21·0	7	28	
	332	0·873	115·5					331
		0·860	65·3					344
		0·967	74·8			(255)	(959)	(2085)
June 19								
170·766	330	0·377	10·8	233·5	+23·6	29	89	
	331	0·493	38·5	218·3	+24·4	0	9	
	331	0·520	42·0	215·5	+24·4	5	21	
	331	0·567	43·8	211·2	+25·8	2	19	
	331	0·580	44·3	211·2	+26·1	5	18	
	332	0·660	119·1	200·9	−17·2	0	15	
	332	0·681	120·3	199·7	−18·6	56	316	
	332	0·673	119·3	203·4	−23·7	81	248	
	332	0·705	128·8	200·9	−24·6	8	22	
	332	0·726	125·4	198·0	−25·3	8	37	
	332	0·729	128·1	198·7	−25·2	11	42	
	332	0·748	127·2	196·8	−25·4	3	17	
	332	0·763	120·1	193·0	−21·1	0	10	
	332	0·770	116·4	191·3	−18·7	6	31	
	332	0·790	118·3	190·0	−20·7	3	17	
		0·929	73·9			(217)	(912)	567 (567)
June 20*								
171·969		0·930	244·7					148
		0·913	309·8					99
	330	0·405	334·7	232·8	+23·4	19	106	
	331	0·385	14·4	216·0	+23·8	15	26	
	331	0·436	22·8	211·2	+25·6	28	45	
	332	0·414	144·0	207·2	−17·5	0	10	
	332	0·414	135·5	200·9	−18·4	60	281	
	332	0·505	141·4	201·3	−21·2	0	4	
	332	0·519	147·3	203·7	−24·4	71	368	
	332	0·553	141·0	199·8	−23·5	0	35	
	332	0·583	142·2	198·7	−25·4	25	47	
	331	0·610	140·3	196·3	−26·0	16	51	
	332	0·551	126·0	194·3	−17·0	0	7	
	332	0·656	124·3	186·9	−19·9	0	14	
		0·759						336
		0·871				(234)	(994)	862 (1445)
June 21								

Mean Solar Time.	No. of Group and Letter for Spot.	Distance from Centre in terms of Sun's Radius.	Position Angle from Sun's Axis.	Heliographic Longitude.	Heliographic Latitude.
1880. 172ᵈ·734	330	0·495	317·7	233·0	+23·3
	331	0·379	347·8	216·8	+23·8
	331	0·405	2·5	210·7	+25·9
	332	0·468	163·4	203·4	−24·5
	332	0·387	152·7	201·1	−18·0
	332	0·485	152·7	197·8	−23·4
	332	0·506	155·7	198·5	−25·3
	332	0·514	152·6	196·7	−25·0
	332	0·531	152·4	196·0	−25·9
	332	0·545	151·9	195·2	−26·6
June 22*					
173·964	330	0·669	303·7	232·8	+23·6
	331	0·462	331·5	209·7	+26·1
	331	0·464	341·3	205·2	+28·2
	332	0·466	193·6	202·4	−24·6
	332	0·353	194·9	201·0	−17·6
	332	0·390	190·2	199·7	−20·2
	332	0·433	186·4	198·6	−23·7
	332	0·382	184·8	197·5	−20·1
	332	0·470	180·8	195·9	−25·6
	332	0·468	177·0	194·0	−25·5
	332	0·418	165·8	189·3	−21·6
June 23					
175·027	332	0·541	227·2	206·3	−19·3
	332	0·461	223·3	200·7	−17·3
	332	0·488	219·6	200·7	−19·8
	332	0·555	215·3	202·0	−24·6
	332	0·507	210·1	197·5	−23·6
	332	0·460	212·9	196·9	−20·3
	332	0·512	203·4	194·4	−25·6
June 24					
175·601	332	0·624	223·8	202·0	−24·4
	332	0·571	228·8	200·9	−19·8
	332	0·545	231·9	200·4	−17·4
	332	0·568	219·8	197·1	−23·4
	332·	0·523	223·1	196·1	−20·1
	332	0·563	213·2	193·7	−25·6
June 25†					
177·630		0·915	298·3		
	332	0·879	246·3	205·3	−19·2
	332	0·851	240·2	200·3	−23·3
	332	0·849	245·7	201·7	−18·9
	332	0·828	247·8	199·9	−16·6
	332	0·811	243·1	197·1	−19·8
	332	0·809	234·6	194·0	−26·0
	332	0·803	236·7	194·4	−24·3
	332	0·796	240·4	194·8	−21·3
	332	0·786	243·0	194·6	−19·0
	333	0·946	119·2	80·0	−26·4
June 27*					

* Indian photo.   The Areas of Spots and Faculæ are expressed in Millionths of the Sun's visible Hemisphere.

MEASURES of POSITIONS and AREAS of SPOTS and FACULÆ upon the SUN'S DISK on PHOTOGRAPHS taken in the YEAR 1880—*continued.*

Mean Solar Time	No. of Group, and Letter for Spot	Distance from Centre in terms of Sun's Radius	Position Angle from Sun's Axis	HELIOGRAPHIC Longitude	HELIOGRAPHIC Latitude	SPOTS. Area of UMBRA for each Spot (and for Day)	SPOTS. Area of WHOLE Spot (and for Day)	FACULÆ. Area for each Group (and for Day)
1880.								
179d·011		0·863	295·2					297
	332	0·964	244·9	200·3	−23·2	70	331	1083 f
	332	0·958	251·1	200·1	−17·2	30	92	
	332	0·939	240·1	194·2	−26·7	0	33	
	333	0·826	124·8	79·8	−26·2	7	12	
	333	0·908	133·1	73·2	−36·7	11	81	
	333	0·937	133·4	68·3	−38·7	0	34	
		0·865	59·5					142
June 28						(118)	(583)	(2054)
180·086		0·938	294·3					369
		0·867	285·2					543
	332	0·992	243·0	194·7	−26·3	0	3	1058 f
	334	0·462	222·0	133·3	−17·2	28	148	
	334	0·446	214·9	130·1	−18·5	8	46	
	335	0·444	321·5	131·9	+23·1	3	12	
	335	0·431	326·0	129·7	+23·7	0	8	
	333	0·826	138·6	72·2	−36·2	11	30	412 n p
	333	0·878	137·9	65·8	−38·7	18	40	
June 29						(68)	(287)	(2382)
181·045		0·988	243·2					698
		0·900	286·0					371
	334	0·604	235·9	133·2	−17·1	41	158	
	334	0·576	233·9	130·8	−17·1	0	12	
	334	0·587	230·6	130·4	−19·1	0	7	
	334	0·565	230·8	129·1	−18·1	7	23	
	333	0·583	145·6	80·4	−25·8	0	11	
	333	0·743	147·3	72·2	−35·9	9	27	
	333	0·811	144·2	64·5	−38·6	8	17	
June 30						(65)	(255)	(1069)
182·902		0·965	293·2					608
	●	0·828	296·2					109
	334	0·853	247·6	132·6	−17·1	39	185	319 s f
	333	0·627	172·9	71·8	−35·1	6	30	
	333	0·659	170·2	69·1	−37·1	5	21	
	333	0·676	162·8	62·8	−36·9	6	11	
July 2						(56)	(247)	(1036)
183·997	334	0·954	250·6	133·0	−17·4	41	206	355 s f
	333	0·648	191·2	72·4	−35·8	2	6	
	333	0·649	188·6	69·6	−36·4	10	65	
	333	0·666	187·0	68·6	−37·9	7	19	
	333	0·651	179·2	62·1	−37·8	7	9	
July 3						(67)	(305)	(355)
184·871	334	0·992	251·9	132·2	−17·4	57	190	720 c
	333	0·689	202·0	69·8	−36·3	31	86	
	333	0·698	198·2	67·2	−38·0	8	53	
	336	0·688	122·6	13·5	−19·0	11	42	
		0·969	112·6					421
July 4*						(107)	(351)	(1141)

Mean Solar Time	No. of Group, and Letter for Spot	Distance from Centre in terms of Sun's Radius	Position Angle from Sun's Axis	HELIOGRAPHIC Longitude	HELIOGRAPHIC Latitude	SPOTS. Area of UMBRA for each Spot (and for Day)	SPOTS. Area of WHOLE Spot (and for Day)	FACULÆ. Area for each Group (and for Day)
1880.								
185d·938	333	0·777	215·7	71·0	−36·1	27	86	462 p
	333	0·766	210·2	66·3	−38·2	2	15	
	336	0·508	140·3	17·0	−19·6	12	32	
	336	0·525	136·6	14·7	−19·0	0	10	
	336	0·551	135·6	12·9	−19·8	17	53	
		0·888	116·0	●				335
July 5						(58)	(196)	(797)
186·631	333	0·834	221·5	70·8	−35·9	0	45	
	336	0·436	157·0	17·5	−20·0	0	31	
	336	0·475	148·0	12·4	−20·1	21	73	
	336	0·467	146·1	12·0	−19·2	0	22	
		0·907	118·4					1854
		0·964	75·3					3366
July 6†						(21)	(171)	(5220)
188·903		0·945	287·0					101
	333	0·980	231·8	70·2	−36·2	38	190	
	333	0·961	228·0	63·5	−38·5	61	245	
	336	0·526	218·1	18·0	−20·7	19	80	
	336	0·496	215·1	15·4	−20·2	2	6	
	336	0·479	208·7	12·1	−21·0	24	82	
		0·808	121·8					118
		0·941	112·3					269
July 8						(144)	(603)	(1210)
189·980	333	0·995	230·5	62·7	−38·6	59	230	
	336	0·677	233·3	18·8	−20·5	8	29	
	336	0·657	231·7	16·9	−20·6	0	7	
	336	0·630	229·6	14·2	−20·6	0	6	
	336	0·608	227·3	12·0	−20·7	15	56	
		0·743	131·3					226
		0·845	117·0					227
July 9		0·946	111·3					261
						(82)	(328)	(1100)
190·897	336	0·798	240·5	19·0	−20·4	6	9	
	336	0·788	239·4	17·8	−20·8	0	8	
		0·745	124·3					230
July 10		0·875	114·3					204
						(6)	(17)	(434)
191·638	336	0·882	243·6	19·1	−20·8	3	10	290 c
July 11*						(3)	(10)	(290)
192·891		0·955	246·0					621
July 12*								(621)
193·922	337	0·799	240·3	305·3	−34·0	0	26	154
		0·652	197·8					
		0·885	65·3					88
July 13						(0)	(26)	(242)
194·897		0·967	111·8					275
July 14								(275)

* Indian photo.   The Areas of Spots and Faculæ are expressed in Millionths of the Sun's visible Hemisphere.   † Melbourne photo.

MEASURES of POSITIONS and AREAS of SPOTS and FACULÆ upon the SUN'S DISK on PHOTOGRAPHS taken in the YEAR 1880—*continued.*

Mean Solar Time.	No. of Group and Letter for Spot.	Distance from Centre in terms of Sun's Radius.	Position Angle from Sun's Axis.	HELIOGRAPHIC Longitude.	HELIOGRAPHIC Latitude.	SPOTS Area of UMBRA for each Spot (and for Day).	SPOTS Area of WHOLE Spot (and for Day).	FACULÆ Area for each Group (and for Day).
1880. 195ᵈ·951 July 15		0·905	115·8					419 (419)
197·538 July 17†		0·829	118·8					1840
		0·874	77·3					1271 (3111)
200·113 July 19		0·914	246·5					358
		0·868	64·3					406 (764)
201·087	338	0·306	28·0	187·8	+20·6	14	76	
	338	0·406	47·8	177·9	+20·6	17	123	
		0·975	250·3					160
		0·894	233·2					184
		0·822	60·2					404
		0·930	114·0					130
July 20						(31)	(199)	(878)
202·007	338	0·269	339·2	190·2	+19·6	4	31	
	338	0·288	350·7	187·3	+21·6	17	54	
	338	0·298	10·3	181·1	+22·1	39	154	
	338	0·281	14·9	180·7	+18·5	0	32	
	338	0·270	15·3	180·1	+20·2	4	5	
	338	0·338	20·7	177·0	+23·4	10	72	
	338	0·295	28·6	175·8	+20·0	0	4	
	339	0·970	67·5	107·8	+23·5	0	33	284 c
		0·810	51·8					123
		0·844	118·2					97
July 21						(74)	(385)	(504)
202·719	338	0·371	312·6	191·8	+19·5	1	11	
	338	0·355	314·7	190·5	+19·5	0	20	
	338	0·360	321·3	188·9	+21·4	4	18	
	338	0·344	324·4	187·3	+21·2	12	46	
	338	0·251	337·1	180·9	+18·5	0	45	
	338	0·294	342·0	180·6	+21·4	48	258	
	338	0·312	351·8	177·8	+23·1	17	78	
	339	0·307	356·4	176·2	+23·0	3	25	
	338	0·362	356·7	176·3	+26·3	0	9	
	339	0·923	67·3	107·7	+22·9	11	32	631 c
		0·975	127·6					229
July 22*						(86)	(542)	(860)
204·073		0·787	240·0					1297
	338	0·818	293·1	211·0	+21·9	0	8	313 p
	338	0·789	296·2	207·5	+23·8	0	8	
	338	0·553	301·6	187·4	+21·4	0	6	
	338	0·462	301·9	181·5	+18·9	0	15	
	338	0·490	308·7	181·5	+22·6	65	290	
	338	0·444	306·0	179·5	+20·0	4	18	
	338	0·418	299·7	179·3	+16·8	0	9	
	338	0·454	314·7	177·6	+23·5	32	194	
	339	0·422	312·1	176·7	+21·4	0	6	
	338	0·406	310·2	176·3	+20·8	0	9	
	339	0·781	64·7	107·2	+22·9	0	7	444 nf
		0·895	134·7					280
July 23						(101)	(570)	(2334)

Mean Solar Time.	No. of Group and Letter for Spot.	Distance from Centre in terms of Sun's Radius.	Position Angle from Sun's Axis.	HELIOGRAPHIC Longitude.	HELIOGRAPHIC Latitude.	SPOTS Area of UMBRA for each Spot (and for Day).	SPOTS Area of WHOLE Spot (and for Day).	FACULÆ Area for each Group (and for Day).
1880. 205ᵈ·083		0·875	240·8					1296
	338	0·934	290·8	213·1	+21·3	0	14	242 f
	338	0·933	292·5	212·8	+22·9	0	27	
	338	0·653	299·4	181·7	+22·9	49	231	
	338	0·615	297·2	180·1	+20·9	6	22	
	338	0·606	301·5	177·7	+22·9	8	67	
	338	0·591	303·5	176·1	+23·5	17	86	
		0·834	141·2					83
July 24						(80)	(447)	(1621)
207·053		0·951	242·8					811
	338	0·902	293·5	181·9	+23·5	32	208	1157 c
	338	0·886	291·7	179·8	+21·7	0	9	
	338	0·866	290·7	177·4	+20·7	6	22	
	338	0·846	294·3	174·6	+23·4	4	28	
	340	0·847	64·3	60·8	+24·6	0	10	338 e
	340	0·857	63·2	59·3	+25·7	5	20	
	340	0·879	62·5	57·1	+26·7	0	2	
	340	0·890	64·3	55·4	+25·3	0	8	
	341	0·962	78·0	43·0	+13·1	0	8	360 sf
		0·948	132·7					241
July 26						(47)	(315)	(2907)
207·946		0·979	244·0					559
	338	0·968	293·1	182·2	+23·7	29	187	747 f
	338	0·924	293·7	173·4	+24·6	0	42	
	340	0·750	61·0	59·4	+25·2	5	32	365 e
	340	0·799	62·2	54·6	+25·4	0	14	
	341	0·884	78·6	43·4	+12·7	4	20	263 e
	341	0·916	77·8	39·1	+13·4	2	10	
		0·920	135·0					380
July 27						(40)	(305)	(2314)
208·590	340	0·665	37·6	59·1	+25·3	0	66	
	341	0·831	77·2	41·0	+13·8	37	192	
		0·850	140·8					782
		0·923	134·9					856
July 28†						(37)	(258)	(1638)
210·026		0·953	294·8					536
		0·881	241·7					115
		0·736	210·7					61
	340	0·448	41·7	59·5	+25·1	10	30	
	341	0·615	76·4	40·0	+13·0	9	38	
		0·752	154·2					221
		0·863	143·3					327
		0·904	117·0					379
		0·974	109·2					169
July 29						(19)	(68)	(1808)
211·009		0·948	246·0					47
		0·806	213·5					66
		0·751	305·5					382
	340	0·33g	14·8	59·9	+24·9	5	37	
		0·775	155·7					700
		0·850	121·7					555
July 30						(5)	(37)	(1750)

* Indian photo.   The Areas of Spots and Faculæ are expressed in Millionths of the Sun's visible Hemisphere.   † Melbourne photo.

MEASURES of POSITIONS and AREAS of SPOTS and FACULÆ upon the SUN'S DISK on PHOTOGRAPHS taken in the YEAR 1880—*continued.*

Left half

Mean Solar Time	No. of Group and Letter for Spot	Distance from Centre in terms of Sun's Radius	Position Angle from Sun's Axis	HELIOGRAPHIC		SPOTS		FACULÆ
				Longitude	Latitude	Area of UMBRA for each Spot (and for Day)	Area of WHOLE Spot (and for Day)	Area for each Group (and for Day)
1880. 211d.930	340	0·823	297·5	59·8	+24·6	5	19	267
		0·339	341·9					
		0·757	160·7					356
		0·776	122·7					377
		0·931	108·2					8g
July 31						(5)	(19)	(1089)
213·795	342	0·882	79·6	326·2	+12·0	51	485	
	342	0·91	79·6	312·0	+12·0	0	5	
	342	0·921	79·4	320·5	+12·1	0	15	
	342	0·933	80·8	319·1	+10·7	46	257	530 c
Aug. 2*						(97)	(762)	(530)
215·093	342	0·703	82·5	326·5	+ 9·6	0	2	162 f
	342	0·708	79·0	326·2	+12·1	104	455	
	342	0·710	82·1	325·2	+ 9·9	0	11	
	342	0·735	80·1	313·9	+11·4	0	9	
	342	0·758	77·8	322·0	+13·2	0	3	
	342	0·760	81·2	321·7	+10·7	5	34	
	342	0·792	79·8	320·0	+11·8	10	24	
	342	0·792	82·0	318·8	+10·1	16	150	
	342	0·801	80·2	317·9	+11·5	0	24	
	342	0·874	79·4	300·1	+12·2	0	15	
Aug. 3						(135)	(727)	(162)
215·735	342	0·600	77·7	326·1	+12·3	93	411	
	342	0·629	79·1	323·9	+11·6	8	56	
	342	0·657	80·4	321·7	+10·9	16	54	
	342	0·679	79·0	320·0	+11·9	6	19	
	342	0·696	81·6	318·6	+10·3	32	158	
	342	0·779	78·4	311·5	+12·8	15	30	
	342	0·850	77·6	304·3	+13·7	2	12	
Aug. 4*						(172)	(740)	
216·694		0·891	223·9					470
	342	0·423	74·5	325·7	+12·1	102	499	
	342	0·480	78·9	321·7	+10·8	7	35	
	342	0·507	77·4	320·0	+11·7	0	17	
	342	0·513	79·1	319·3	+10·9	2	11	
	342	0·516	80·9	318·5	+10·0	35	161	
	342	0·528	78·6	318·5	+11·3	5	24	
	342	0·604	77·5	313·2	+12·5	14	59	
Aug. 5*						(165)	(806)	(470)
217·594		0·916	223·6					1030
		0·887	279·8					801
	342	0·196	70·2	327·4	+10·0	0	5	
	342	0·215	70·7	326·4	+10·2	0	7	
	342	0·246	64·4	325·1	+12·2	77	381	
	342	0·299	73·2	311·3	+10·9	0	24	
	342	0·348	76·8	318·1	+10·5	32	133	
	342	0·420	73·8	313·9	+12·5	11	24	
	342	0·445	73·5	312·4	+12·9	0	35	
Aug. 6†						(120)	(509)	(1831)

Right half

Mean Solar Time	No. of Group and Letter for Spot	Distance from Centre in terms of Sun's Radius	Position Angle from Sun's Axis	HELIOGRAPHIC		SPOTS		FACULÆ
				Longitude	Latitude	Area of UMBRA for each Spot (and for Day)	Area of WHOLE Spot (and for Day)	Area for each Group (and for Day)
1880. 218d.649		0·960	224·7					238
		0·869	241·5					267
	342	0·104	351·5	325·2	+12·2	130	490	
	342	0·082	11·6	323·3	+10·8	4	14	
	342	0·079	21·4	322·7	+10·5	0	10	
	342	0·099	28·2	321·6	+11·3	0	14	
	342	0·122	30·5	320·7	+12·3	0	17	
	342	0·128	44·7	319·1	+11·5	14	72	
	342	0·189	57·0	315·0	+12·1	4	13	
	343	0·999	66·5	235·1	+23·6	0	83	
Aug. 7*						(152)	(713)	
219·734	343a	0·899	247·7	10·1	−16·8	3	16	86 c
	343a	0·884	247·0	8·1	−16·9	9	33	
	343b	0·560	212·9	329·1	−21·9	9	27	
	342	0·574	291·5	325·0	+11·9	124	447	
	342	0·190	300·3	319·6	+11·8	2	21	
	342	0·170	295·9	318·9	+10·6	16	58	
	342	0·123	338·9	311·6	+13·0	0	20	
	342	0·128	346·5	311·7	+13·4	0	10	
	342	0·125	6·5	309·2	+13·6	2	17	
	343	0·955	66·9	236·3	+23·8	48	292	642 s f
Aug. 8*						(210)	(941)	(728)
221·082		0·900	249·3					230
	342	0·540	281·8	324·7	+11·8	91	481	
	342	0·466	283·2	319·7	+11·8	0	6	
	342	0·454	280·2	319·1	+10·4	6	16	
	342	0·393	286·6	314·7	+12·4	10	24	
	342	0·376	287·4	313·6	+12·5	0	7	
	342	0·347	291·2	311·5	+13·3	0	2	
	342	0·309	295·9	308·7	+13·9	13	30	
	343	0·838	65·7	135·8	+23·8	44	250	741 s f
	344	0·942	111·3	225·1	−17·5	44	150	127 c
	345	0·976	69·3	213·5	+21·6	45	166	215 n
	346	0·981	111·2	216·7	−19·2	0	48	103 p
Aug. 9						(253)	(1180)	(1416)
221·907	342	0·785	235·7					141
	342	0·684	280·5	324·5	+11·9	111	461	
	342	0·613	278·4	319·2	+10·3	0	1	
	342	0·571	281·5	316·0	+11·9	11	22	
	342	0·550	283·3	314·4	+12·7	0	2	
	342	0·513	285·1	311·7	+13·3	2	11	
	342	0·470	287·3	308·7	+13·8	14	30	
	342	0·741	63·3	235·0	+24·0	104	329	485 s f
	344	0·867	114·4	225·7	−17·4	27	158	325 c
	344	0·986	114·2	222·1	−18·3	0	11	
	345	0·923	69·6	213·0	+21·3	28	239	425 u f
	346	0·934	113·5	216·4	−19·1	8	70	381 e
Aug. 10						(305)	(1334)	(1757)

* Indian photo.   The Areas of Spots and Faculæ are expressed in Millionths of the Sun's visible Hemisphere.   † Melbourne photo.

MEASURES of POSITIONS and AREAS of SPOTS and FACULÆ upon the SUN'S DISK on PHOTOGRAPHS taken in the YEAR 1880—*continued.*

Mean Solar Time.	No. of Group, and Letter for Spot.	Distance from Centre in terms of Sun's Radius.	Position Angle from Sun's Axis.	HELIOGRAPHIC Longitude.	HELIOGRAPHIC Latitude.	SPOTS. Area of UMBRA for each Spot (and for Day).	SPOTS. Area of WHOLE for each Spot (and for Day).	FACULÆ. Area for each Group (and for Day).
1880. 222d·969	342	0·839	280·4	324·6	+12·3	89	443	392 c
	342	0·761	279·6	317·0	+11·5	0	13	
	342	0·665	284·2	308·7	+14·3	0	10	
	343	0·585	56·5	235·0	+24·4	45	425	
	344	0·738	120·6	225·7	−17·1	16	79	
	345	0·813	69·0	2·3·1	+20·9	38	245	316 nf
	346	0·835	118·4	216·3	−19·3	16	57	249 f
Aug. 11						(204)	(1272)	(957)
223·985		0·967	245·0					103
		0·749	235·0					67
	342	0·940	281·0	324·5	+12·6	106	439	977 nf
	342	0·898	279·2	318·2	+11·2	3	20	
	343	0·436	43·6	234·5	+24·6	110	444	
	344	0·596	130·6	225·6	−16·9	24	131	
	345	0·665	65·6	213·5	+21·0	60	264	267 nf
	346	0·705	125·4	216·5	−18·8	15	66	
	346	0·726	124·7	214·7	−19·3	2	24	
		0·916	70·7					519
Aug. 12						(320)	(1388)	(2568)
224·988	342	0·994	281·6	325·3	+12·2	0	279	634 nf
	343	0·326	17·7	234·3	+24·6	106	364	
	344	0·464	147·3	225·4	−16·5	27·	129	
	345	0·501	58·8	213·3	+21·0	46	211	
	346	0·364	140·7	218·3	−16·9	0	7	
	346	0·574	137·4	216·4	−18·9	21	122	
	346	0·617	136·1	213·5	−20·4	1	4	
	346	0·613	132·5	212·2	−18·6	5	12	
	346	0·649	133·9	210·6	−20·9	6	20	
		0·862	68·5					630
		0·911	123·2					680
Aug. 13						(212)	(1148)	(1944)
227·618	347	0·757	291·9	254·4	+20·9	10	43	
	347	0·721	294·2	250·8	+22·0	0	42	
	343	0·548	306·3	234·8	+24·9	43	290	
	344	0·509	220·9	226·0	−16·2	19	91	
	345	0·280	335·2	213·0	+21·4	29	176	
	346	0·453	203·4	216·7	−17·8	18	90	
	346	0·432	200·7	215·0	−17·1	9	51	
	346	0·451	199·1	214·7	−18·4	0	11	
	346	0·443	189·5	210·2	−19·0	0	26	
	346	0·465	187·1	209·3	−20·6	0	35	
	346	0·467	183·3	207·4	−20·9	13	39	
Aug. 16†						(141)	(894)	
229·052	347	0·933	289·5	256·4	+20·6	31	76	215 nf
	343	0·764	297·6	235·0	+25·3	39	217	764 c
	344	0·719	239·4	226·7	−16·2	32	106	272 s
	345	0·486	303·8	212·4	+21·8	44	203	
	346	0·593	224·4	212·6	−13·9	47	227	
Aug. 17						(193)	(829)	(1251)
1880. 229d·741	347	0·979	288·9	257·4	+19·9	20	105	385 f
	343	0·848	295·0	235·3	+24·8	35	227	616 c
	344	0·810	244·0	227·0	−16·3	17	110	
	346	0·714	236·4	216·2	−17·8	4	8	
	346	0·705	238·6	216·4	−16·2	1	5	
	346	0·700	235·0	214·7	−18·1	11	33	
	346	0·685	233·6	213·1	−18·3	22	65	
	346	0·657	230·8	210·1	−18·7	0	5	
	346	0·660	219·5	209·7	−19·5	5	21	
	346	0·657	228·1	209·0	−20·1	21	70	
	346	0·644	224·4	206·5	−21·3	5	23	
	346	0·638	223·5	206·6	−20·4	0	6	
	346	0·633	223·8	205·6	−21·0	1	6	
	345	0·599	297·4	212·5	+21·7	41	204	
	348	0·330	327·5	188·9	+22·8	9	37	
	348	0·315	329·7	187·6	+22·5	0	4	
	348	0·302	321·6	189·2	+20·4	2	6	
	348	0·305	338·0	184·8	+23·1	6	21	
	348	0·300	342·0	183·5	+23·4	4	13	
Aug. 18*						(204)	(969)	(1001)
231·008	343	0·937	294·7	235·2	+25·6	34	133	972 f
	344	0·935	249·8	226·7	−16·1	13	69	144 nf
	346	0·779	293·3	211·5	+22·4	33	203	
	346	0·850	244·2	214·2	−17·0	0	22	589 c
	346	0·834	243·1	212·2	−17·9	0	52	
	346	0·819	240·7	209·9	−19·1	0	25	
	346	0·808	239·2	208·3	−19·7	0	15	
	346	0·788	236·6	205·5	−20·8	10	33	
	348	0·530	303·0	189·0	+22·8	56	316	
	348	0·490	309·3	184·8	+24·3	9	28	
	348	0·467	308·0	183·8	+23·0	6	27	
	348	0·447	304·9	183·4	+21·2	0	5	
	348	0·448	309·4	182·3	+22·9	15	76	
Aug. 19						(176)	(1625)	(1705)
231·718	343	0·987	294·3	234·2	+25·0	25	107	676 nf
	344	0·984	251·6	228·4	−16·6	25	101	242 c
	346	0·909	246·8	212·7	−17·7	0	19	611 c
	345	0·894	244·2	209·9	−19·3	5	21	
	345	0·867	291·3	211·9	+21·9	43	217	
	346	0·659	296·5	191·1	+22·5	67	367	
	348	0·635	301·0	188·3	+24·7	2	7	
	348	0·618	297·3	187·8	+22·1	5	12	
	348	0·593	301·1	185·1	+23·7	10	57	
	348	0·588	298·8	185·3	+22·3	12	48	
	348	0·362	301·1	182·9	+22·8	25	111	
		0·983	68·7					218
Aug. 20*						(219)	(1063)	(1747)
233·003		0·966	245·0					663
		0·829	231·8					593
	345	0·966	291·4	211·0	+22·4	75	303	343 nf
	348	0·842	292·5	191·8	+22·7	75	418	553 c

* Indian photo.   The Areas of Spots and Faculæ are expressed in Millionths of the Sun's visible Hemisphere.   † Melbourne photo.

MEASURES of POSITIONS and AREAS of SPOTS and FACULÆ upon the SUN's DISK on PHOTOGRAPHS taken in the YEAR 1880—*continued.*

Mean Solar Time.	No. of Group, and Letter for Spot.	Distance from Centre in terms of Sun's Radius.	Position Angle from Sun's Axis.	Heliographic Longitude.	Heliographic Latitude.	Area of UMBRA for each Spot (and for Day).	Area of WHOLE Spot (and for Day).	Faculæ Area for each Group (and for Day).
1880.								
233d·003	348	0·784	295·2	185·3	+24·0	3	14	
	348	0·785	294·4	182·7	+21·9	25	134	
	348	0·757	296·4	182·5	+24·4	0	10	549 e
	349	0·891	67·6	71·2	+23·1	19	36	447 sf
Aug. 21						(197)	(915)	(3148)
233·761		0·982	242·6					117
	348	0·925	291·1	193·1	+22·1	86	439	
	348	0·874	292·6	185·7	+23·2	0	50	
	348	0·862	291·2	184·3	+21·8	4	56	
	348	0·846	292·3	182·3	+22·5	14	119	631 c
Aug. 22*						(104)	(664)	(748)
234·705		0·940	233·9					226
	348	0·979	291·6	192·0	+22·5	79	531	
	348	0·947	290·9	184·4	+22·0	13	67	
	348	0·932	290·7	181·8	+21·9	20	82	
	348	0·920	287·3	179·8	+18·6	10	62	
	348	0·908	289·0	177·7	+20·2	17	58	
	348	0·892	287·8	175·8	+19·0	9	21	834 c
	349	0·702	65·5	68·7	+22·0	9	28	
	349	0·728	65·4	66·5	+22·5	3	18	
Aug. 23*						(160)	(867)	(1060)
235·716	348	0·988	290·9	181·7	+21·7	0	36	663 sf
	349	0·531	57·7	69·8	+22·7	21	72	
Aug. 24*						(21)	(108)	(663)
240·006		0·893	304·2					163
	349	0·497	303·0	68·7	+22·1	0	8	
	349	0·457	305·6	65·6	+22·0	0	14	
	350	0·564	114·3	61·9	−20·9	6	13	
	350	0·527	211·6	59·0	−19·7	10	23	
	351	0·958	80·1	327·9	+11·5	0	9	
	351	0·965	82·1	326·6	+ 9·5	49	180	204 nf
		0·923	147·0					164
		0·937	72·0					98
		0·952	115·0					147
Aug. 28						(65)	(247)	(776)
240·736	349	0·636	295·9	67·1	+21·8	11	47	
	349	0·574	298·1	65·1	+21·8	16	62	
	351	0·905	82·6	326·8	+ 9·7	17	90	
Aug. 29*						(44)	(197)	
241·785	349	0·784	291·6	69·9	+21·4	35	173	
	349	0·720	293·5	63·7	+21·8	37	175	
	351	0·778	82·9	327·2	+10·0	24	100	
Aug. 30*						(96)	(448)	
243·016		0·892	242·8					76
	349	0·922	290·3	70·4	+21·5	31	234	956 e
	349	0·865	291·5	62·4	+22·2	43	280	
	351	0·573	82·7	327·2	+10·1	33	155	
		0·766	69·5					464
Aug. 31						(107)	(669)	(1496)

Mean Solar Time.	No. of Group, and Letter for Spot.	Distance from Centre in terms of Sun's Radius.	Position Angle from Sun's Axis.	Heliographic Longitude.	Heliographic Latitude.	Area of UMBRA for each Spot (and for Day).	Area of WHOLE Spot (and for Day).	Faculæ Area for each Group (and for Day).
1880.								
243d·904		0·962	246·0					102
	349	0·982	290·4	71·4	+21·4	31	181	525 e
	349	0·941	291·3	61·8	+22·4	20	262	288 c
	351	0·396	81·5	327·3	+10·0	19	112	
Sept. 1						(70)	(555)	(915)
245·017	349	0·993	293·2	61·3	+23·8	0	218	175 c
	352	0·872	250·0	33·1	−13·3	0	24	
	352	0·854	250·5	31·2	−12·5	0	14	
	352	0·823	249·1	27·7	−12·7	11	31	218 e
	351	0·158	69·2	327·3	+10·4	28	113	
	353	0·969	72·5	258·9	+18·7	0	34	113 nf
Sept. 2						(39)	(434)	(506)
246·059		0·938	221·8					186
		0·923	296·5					70
		0·803	240·3					211
	352	0·927	253·6	27·4	−12·2	0	16	315 p
	351	0·105	300·8	327·4	+10·3	21	105	
	353	0·890	72·9	258·6	+18·5	10	21	511 f
Sept. 3						(31)	(142)	(1293)
246·902		0·984	255·7					196
		0·898	214·8					184
		0·877	244·5					269
	351	0·288	282·7	327·5	+10·6	21	109	
	353	0·792	72·1	258·5	+18·6	8	16	103 f
	354	0·981	65·9	230·2	+25·0	49	134	761 sp
Sept. 4						(78)	(259)	(1832)
247·671	351	0·453	279·0	327·7	+10·5	22	101	
	353	0·674	70·4	259·1	+18·4	5	16	
	355	0·913	112·8	239·2	−17·3	3	13	
	355	0·924	112·6	137·4	−17·7	7	26	403 c
	355	0·948	112·0	233·8	−18·1	24	48	
	354	0·936	66·2	230·4	+24·8	40	148	641 c
	356	0·972	66·8	222·7	+24·2	22	102	
Sept. 5*						(113)	(454)	(1044)
248·781	351	0·659	277·6	327·7	+10·5	24	100	
	353	0·477	64·7	259·3	+18·3	4	18	
	355	0·786	119·5	240·5	−17·7	16	69	
	355	0·809	117·1	237·5	−16·9	0	16	
	355	0·811	119·5	237·8	−18·0	3	9	
	355	0·830	118·2	236·9	−18·1	2	8	
	355	0·851	116·9	233·3	−18·3	38	173	588 nf
	354	0·831	64·4	230·5	+25·2	31	163	327 c
	356	0·890	66·4	222·9	+24·3	9	49	152 f
Sept. 6*								

* Indian photo. The Areas of Spots and Faculæ are expressed in Millionths of the Sun's visible Hemisphere.

MEASURES of POSITIONS and AREAS of SPOTS and FACULÆ upon the SUN'S DISK on PHOTOGRAPHS taken in the YEAR 1880—continued.

Mean Solar Time.	No. of Group, and Letter for Spot.	Distance from Centre in terms of Sun's Radius.	Position Angle from Sun's Axis.	HELIOGRAPHIC Longitude.	Latitude.	SPOTS. Area of UMBRA for each Spot (and for Day).	Area of WHOLE Spot (and for Day).	FACULÆ. Area for each Group (and for Day).
1880. 25ᵈ·083		0'939	290'5					54
		0'897	245'8					247
	351	0'848	278'1	327'5	+10'7	18	105	241 n/
	354	0'662	59'6	230'0	+25'2	32	117	119 e
	355	0'586	134'2	243'0	-17'6	23	68	
	355	0'623	131'1	239'6	-17'8	10	51	
	355	0'693	126'4	233'1	-18'4	16	100	339 e
	356	0'740	64'1	222'5	+23'9	6	19	114 e
	357	0'867	115'4	213'9	-17'7	9	25	793 e
	358	0'880	67'8	207'0	+23'0	13	31	266 n/
	359	0'973	68'6	190'8	+22'4	50	216	241 s p
Sept. 7						(177)	(732)	(2414)
250'980	351	0'937	277'7	327'4	+ 9'7	32	115	289 n/
	354	0'509	51'4	231'3	+25'0	37	130	
	355	0'474	150'9	243'3	-17'4	29	113	
	355	0'470	146'8	241'7	-16'1	0	17	
	355	0'495	147'4	241'0	-17'7	16	77	
	355	0'517	146'4	239'7	-18'6	2	5	
	355	0'529	141'6	237'1	-17'7	0	5	
	355	0'572	137'7	235'4	-18'4	17	89	
	355	0'585	136'4	232'1	-18'5	23	75	
	357	0'681	122'1	220'6	-15'4	3	14	
	357	0'698	121'6	219'2	-15'8	5	19	
	357	0'704	123'6	219'5	-17'2	6	16	
	357	0'710	119'3	217'5	-14'8	13	35	
	357	0'716	121'2	217'7	-16'1	8	28	
	357	0'746	120'2	215'0	-16'7	23	103	
	357	0'770	120'2	213'0	-17'7	3	10	
	358	0'776	65'9	207'0	+23'2	13	40	
	359	0'912	68'0	190'7	+23'0	83	319	
	359	0'950	72'4	184'2	+19'0	26	74	
Sept. 8*						(339)	(1284)	(289)
251'918	355	0'421	178'7	244'2	-17'6	42	206	
	355	0'426	173'5	241'9	-17'7	15	43	
	355	0'445	163'5	237'2	-18'1	6	23	
	355	0'441	160'6	236'0	-17'4	14	54	
	355	0'476	156'1	233'1	-18'7	25	131	
	357	0'528	137'4	223'1	-16'1	25	98	
	357	0'546	132'0	220'6	-14'8	19	82	
	357	0'572	135'7	221'0	-17'6	5	19	
	357	0'582	133'5	218'7	-17'1	20	75	
	357	0'585	131'1	217'6	-16'2	12	48	
	357	0'590	129'4	216'7	-15'6	3	20	
	357	0'618	129'9	215'2	-17'1	27	131	
	354	0'387	35'3	230'5	+25'4	31	137	
	359	0'120	326'7	248'7	+12'9	3	13	
	359	0'098	327'6	247'9	+12'0	6	20	
	358	0'636	61'7	207'4	+23'3	7	21	
	358	0'684	61'8	203'6	+24'3	0	11	
	359	0'810	67'2	191'0	+24'8	78	269	
	359	0'858	72'5	185'3	+18'7	16	57	
Sept. 9*						(354)	(1458)	

Mean Solar Time.	No. of Group, and Letter for Spot.	Distance from Centre in terms of Sun's Radius.	Position Angle from Sun's Axis.	HELIOGRAPHIC Longitude.	Latitude.	SPOTS. Area of UMBRA for each Spot (and for Day).	Area of WHOLE Spot (and for Day).	FACULÆ. Area for each Group (and for Day).
1880. 253ᵈ·100	355	0'917	280'2					64
	355	0'481	210'6	244'0	-17'4	94	338	
	355	0'450	206'6	241'2	-16'6	10	33	
	355	0'453	203'4	240'0	-17'4	8	19	
	355	0'460	202'2	239'7	-18'0	0	8	
	355	0'466	199'4	238'5	-18'8	4	6	
	355	0'447	198'6	237'6	-17'8	5	30	
	355	0'424	197'8	236'9	-16'5	7	42	
	357	0'427	193'3	235'1	-17'3	53	153	
	355	0'456	188'6	233'3	-19'5	2	14	
	355	0'439	187'2	232'5	-18'5	11	31	
	357	0'414	172'9	226'1	-16'9	3	15	
	357	0'405	169'9	224'9	-16'2	80	259	
	357	0'441	160'8	220'4	-17'4	19	54	
	357	0'410	159'8	220'8	-15'4	2	13	
	357	0'395	158'7	220'7	-14'3	9	53	
	357	0'428	158'9	220'0	-16'3	3	21	
	357	0'453	156'5	218'3	-17'3	22	70	
	357	0'417	155'9	219'0	-15'2	15	43	
	357	0'422	152'3	217'5	-14'8	4	14	
	357	0'458	154'2	217'2	-17'2	0	11	
	357	0'444	152'1	216'5	-17'3	0	7	
	357	0'442	152'1	216'6	-16'1	13	58	
	357	0'476	150'4	215'0	-17'3	11	24	
	357	0'482	148'8	214'0	-17'3	15	52	
	354	0'335	353'3	231'7	+26'6	3	16	
	354	0'308	356'6	230'3	+25'1	20	76	
	356	0'256	4'5	227'9	+22'0	0	4	
	356	0'286	15'0	224'6	+23'2	0	4	
	356	0'301	19'4	222'9	+23'6	0	3	
	358	0'456	50'8	206'6	+23'4	5	13	
	359	0'647	63'8	190'5	+22'2	67	218	
	359	0'702	70'8	185'9	+18'6	14	47	
	359	0'771	68'9	179'1	+20'8	0	10	595 f
	359	0'865	127'3					553
Sept. 10						(499)	(1759)	(1212)
253'766	355	0'563	222'9	243'9	-17'7	123	480	
	355	0'536	221'3	241'9	-17'0	12	46	
	355	0'529	219'0	240'6	-17'4	0	6	
	355	0'515	217'5	239'8	-17'8	14	37	
	355	0'516	215'9	239'6	-18'9	2	9	
	355	0'526	212'4	230'6	-19'4	2	8	
	357	0'515	214'8	238'3	-18'1	0	9	
	355	0'504	213'1	237'1	-18'0	8	45	
	357	0'489	212'7	236'4	-17'3	5	11	
	357	0'487	214'6	237'0	-16'6	13	46	
	355	0'478	210'2	234'9	-17'4	35	164	
	355	0'509	208'3	235'1	-19'6	0	16	
	355	0'500	204'7	233'1	-19'9	7	18	
	355	0'483	204'6	232'5	-18'3	18	40	
	357	0'422	194'3	226'5	-16'9	69	235	
	357	0'405	190'1	224'5	-16'2	17	60	
	357	0'425	188'5	224'1	-17'5	4	15	
	357	0'389	189'0	223'9	-15'3	15	37	

* Indian photo.   The Areas of Spots and Faculæ are expressed in Millionths of the Sun's visible Hemisphere.

MEASURES of POSITIONS and AREAS of SPOTS and FACULÆ upon the SUN'S DISK on PHOTOGRAPHS taken in the YEAR 1880—*continued.*

Mean Solar Time.	No. of Group, and Letter for Spot.	Distance from Centre in terms of Sun's Radius.	Position Angle from Sun's Axis.	Longi-tude.	Lati-tude.	Area of UMBRA for each Spot (and for Day).	Area of WHOLE Spot (and for Day).	Area for each Group (and for Day).
1880.								
53ʰ·766	357	0·418	184·3	222·2	−17·3	2	13	·
	357	0·407	181·6	211·0	−16·6	3	15	
	357	0·404	179·5	220·1	−16·4	15	39	
	357	0·376	181·4	220·7	−14·8	23	154	
	357	0·423	174·1	217·7	−17·6	25	125	
	357	0·401	171·8	216·9	−16·1	52	190	
	357	0·441	167·0	214·5	−17·6	24	86	
	357	0·439	164·7	213·2	−18·4	0	6	
	354	0·378	334·5	230·8	+26·9	2	5	
	354	0·346	331·5	230·7	+24·7	32	71	
	358	0·353	37·0	207·0	+23·3	0	7	
	358	0·389	34·7	206·2	+25·5	0	13	
	359	0·534	58·9	190·8	+22·3	67	209	
	359	0·588	68·0	185·4	+18·7	18	35	
						(607)	(2250)	
Sept. 11*								
254·821	355	0·697	235·2	243·3	−17·6	102	469	
	355	0·707	232·0	242·7	−20·0	0	11	
	355	0·667	232·3	240·1	−18·0	43	152	
	355	0·650	228·5	237·4	−19·3	16	64	
	355	0·631	229·7	236·8	−17·9	12	41	
	355	0·636	226·4	235·7	−19·7	18	70	
	355	0·604	227·9	234·4	−17·5	40	137	
	355	0·596	223·4	232·1	−19·1	5	18	
	357	0·317	219·4	226·5	−16·8	119	453	
	357	0·474	214·9	222·8	−15·9	3	9	
	357	0·446	213·2	222·1	−14·9	25	102	
	357	0·487	209·5	221·0	−18·0	9	36	
	357	0·432	208·1	218·6	−15·3	16	65	
	357	0·452	206·4	218·6	−16·8	14	46	
	357	0·469	201·6	217·4	−18·5	1	10	
	357	0·425	203·4	216·6	−15·7	31	166	
	357	0·450	201·9	216·6	−17·5	22	64	
	357	0·437	200·9	215·5	−17·1	11	35	
	357	0·436	195·9	213·7	−17·5	16	50	
	354	0·482	309·8	230·4	+24·5	11	42	
	354	0·478	312·8	229·3	+25·5	5	18	
	359	0·368	44·6	190·4	+22·1	47	222	
						(566)	(2280)	
Sept. 12*								
256·108		0·844	289·8					381
	355	0·850	244·5	242·7	−17·2	130	652	
	355	0·849	241·7	241·6	−19·4	0	9	
	355	0·799	240·7	236·4	−18·1	8	20	
	355	0·792	237·6	234·7	−20·0	12	97	
	355	0·773	240·7	234·1	−17·1	12	88	
	355	0·775	239·7	233·9	−17·9	0	5	
	355	0·771	238·9	233·3	−18·3	6	9	
	355	0·759	235·1	231·7	−19·0	0	8	528 c
	357	0·705	236·9	227·4	−16·9	114	413	
	357	0·676	235·3	224·8	−16·7	5	29	
	357	0·662	235·2	223·8	−16·2	5	19	
	357	0·676	230·7	223·0	−19·3	0	2	
	357	0·634	234·7	221·8	−15·4	19	72	
	357	0·644	230·7	219·9	−17·9	6	15	
1880.								
256ᵈ·108	357	0·600	231·4	218·4	−15·7	18	193	
	357	0·592	227·1	216·3	−17·3	54	256	
	357	0·567	229·6	215·9	−15·1	8	36	
	357	0·563	227·4	214·8	−15·8	3	19	
	357	0·568	223·9	213·7	−17·6	8	44	
	357	0·544	217·4	209·8	−18·8	0	3	
	354	0·673	299·7	229·6	+25·1	0	20	
	358	0·386	312·8	207·2	+22·0	0	7	
	358	0·393	320·1	205·4	+24·4	12	20	
	359	0·261	358·4	189·9	+22·3	44	174	
Sept. 13		0·910	63·2					383
						(484)	(2210)	(1292)
256·748	355	0·905	247·0	241·4	−17·2	126	733	
	355	0·886	245·4	238·5	−17·8	11	96	
	355	0·867	244·2	235·9	−18·1	20	63	
	355	0·861	241·9	234·5	−19·8	0	10	
	355	0·846	244·3	233·7	−17·2	21	55	
	355	0·855	240·8	233·5	−20·4	27	101	
	357	0·778	240·9	226·1	−17·2	122	515	
	357	0·762	240·1	224·5	−17·2	0	7	
	357	0·717	239·7	220·8	−15·7	20	84	
	357	0·725	235·9	220·0	−18·4	0	24	
	357	0·691	237·0	217·9	−16·3	13	56	
	357	0·678	236·5	216·8	−16·2	18	58	
	357	0·674	233·8	215·5	−17·5	49	197	
	357	0·646	235·1	214·2	−15·7	3	11	
	357	0·652	231·4	213·1	−17·9	5	29	
	358	0·491	304·4	206·8	+22·5	5	16	
	358	0·488	309·8	205·1	+24·8	16	56	
	358	0·471	307·4	204·8	+23·2	0	11	
	358	0·475	310·6	204·2	+24·5	6	20	
Sept. 14*		0·293	332·8	189·2	+22·2	42	167	
						(504)	(2309)	
257·741	355	0·985	250·9	245·0	−17·4	64	345	
	355	0·980	249·2	242·9	−18·6	50	300	704 c
	355	0·964	248·9	238·8	−18·1	20	83	
	355	0·946	248·8	235·2	−17·4	6	31	
	355	0·951	242·3	235·3	−20·9	13	63	
	357	0·807	245·8	227·1	−18·0	114	465	
	357	0·842	246·4	220·9	−15·4	22	82	724 c
	357	0·822	243·1	217·8	−17·2	55	256	
	357	0·798	240·8	214·8	−18·0	0	17	
	358	0·655	299·5	206·5	+24·5	15	45	
	359	0·448	306·5	190·7	+21·0	45	186	
	360	0·933	75·2	98·0	+16·4	3	20	136 c
Sept. 15*	361	0·968	73·3	91·0	+18·0	24	68	688 s f
						(431)	(1963)	(2252)
259·090		0·951	295·5					295
	357	0·987	250·5	227·7	−17·9	79	390	
	357	0·975	250·7	224·0	−16·9	0	44	
	357	0·965	251·3	221·8	−15·9	19	51	

* Indian photo.   The Areas of Spots and Faculæ are expressed in Millionths of the Sun's visible Hemisphere.

					1880.		°	°	°			
218.2	−15.9	0	31		263d.714	362	0.994	290.1	172.2	+20.7	46	322	
216.7	−17.2	59	142	918 sf		360	0.295	302.8	163.9	+16.0	14	29	
205.6	+24.4	0	8	209 e		360	0.249	308.5	100.6	+15.7	0	8	
189.0	+22.3	34	184			360	0.224	306.1	99.7	+14.4	0	25	
98.6	+16.4	7	14	104 sf		361	0.189	357.0	89.6	+17.8	7	16	
89.8	+18.0	21	72			364	0.457	59.4	64.4	+19.8	14	41	
88.7	+16.6	0	6	402 nf		364	0.478	63.2	62.3	+18.7	4	11	
				379		364	0.499	62.1	61.2	+19.7	1	6	
		(219)	(942)	(2307)		364	0.515	64.9	59.5	+18.7	9	40	
						364	0.528	60.2	59.7	+21.3	1	7	
				666							(96)	(505)	
				265	Sept. 21*								157
188.8	+22.4	60	213	261 n	265.030		0.909	236.2					
176.0	+21.3	13	27	41 e		360	0.560	288.9	105.0	+16.2	2	7	
99.3	+16.3	7	23			361	0.347	304.6	89.1	+18.0	3	9	
97.4	+15.2	0	8			361	0.337	299.9	88.7	+16.0	0	6	
95.2	+17.1	0	2			361	0.330	304.0	88.2	+17.3	0	4	
89.8	+17.8	21	62	324 f		361	0.343	313.4	87.0	+20.3	0	3	
				170		364	0.246	25.4	65.3	+19.7	14	63	
				85		364	0.267	26.5	64.4	+20.7	0	3	
		(101)	(335)	(1812)		364	0.290	36.1	61.2	+20.4	0	7	
						364	0.271	44.1	60.3	+18.0	0	7	
				238		364	0.289	44.3	59.4	+18.7	2	*15	
				59		364	0.306	43.0	58.9	+19.7	0	13	
				652		366	0.899	116.0	12.7	−19.7	6	42	
188.6	+22.2	81	212	281 c		366	0.907	115.1	11.4	−19.2	0	6	
177.0	+21.5	0	38	231 f		366	0.935	113.9	6.9	−19.4	9	31	463 c
173.0	+20.3	16	57		Sept. 22						(36)	(216)	(620)
101.4	+16.3	12	28										
97.9	+16.0	0	20										
91.4	+17.3	0	16		265.785	361	0.481	293.8	89.0	+17.3	0	18	
91.0	+16.2	0	4			364	0.227	341.5	66.1	+19.3	14	68	
89.8	+15.9	0	6			364	0.205	355.9	62.6	+18.7	0	5	
89.5	+17.6	12	42			364	0.208	11.4	59.2	+18.7	0	10	
89.8	+25.2	0	11			366	0.815	119.8	13.3	−19.3	17	45	
63.6	+19.5	0	44	1103 np	Sept. 23*						(31)	(146)	
59.5	+18.8	0	5										
				52									
				230	266.912	361	0.733	287.6	93.7	+17.5	6	21	
				190		361	0.698	289.8	90.5	+18.7	2	6	
		(121)	(483)	(3046)		364	0.392	304.1	66.8	+19.2	17	64	
				304		364	0.324	312.9	61.3	+19.4	0	16	
				231		364	0.313	315.7	60.2	+19.6	0	23	
				568 n		366	0.661	128.8	14.0	−18.7	5	28	
175.3	+21.7	29	211			366	0.679	127.3	12.2	−18.6	0	7	340 c
100.3	+14.2	0	1			368	0.961	110.9	336.4	−17.8	36	115	302
102.8	+16.4	12	30				0.868	118.3					
89.4	+17.9	11	22		Sept. 24*						(66)	(280)	(642)
64.9	+20.0	0	7										
63.2	+19.5	0	8		267.936		0.877	289.5						607
61.7	+19.2	0	1				0.728	232.0					107
59.4	+18.9	6	13			364	0.575	293.7	67.0	+19.1	14	38	
53.3	+23.7	0	18	222 e		364	0.557	294.1	65.7	+18.9	0	6	
				55		364	0.526	297.2	63.0	+19.9	1	30	
				110		365	0.473	305.8	57.7	+22.3	0	5	
				400		366	0.524	144.9	14.8	−18.8	5	16	35 f
		(58)	(311)	(1890)		367	0.756	125.6	352.3	−20.9	0	16	217 f

The Areas of Spots and Faculæ are expressed in Millionths of the Sun's visible Hemisphere.

MEASURES of POSITIONS and AREAS of SPOTS and FACULÆ upon the Sun's DISK on PHOTOGRAPHS taken in the YEAR 1880—*continued*.

Mean Solar Time.	No. of Group, and Letter for Spot.	Distance from Centre in terms of Sun's Radius.	Position Angle from Sun's Axis.	HELIOGRAPHIC Longitude.	Latitude.	SPOTS. Area of UMBRA for each Spot (and for Day).	Area of WHOLE Spot for each Spot (and for Day).	FACULÆ. Area for each Group (and for Day).
1880. 267d·936	368	0·875	115·7	337·2	−18·6	47	264	
	369	0·909	115·0	332·7	−19·3	0	17	
	369	0·935	164·4	328·6	−19·8	0	40	503 c
		0·927	76·5					655
Sept. 25						(67)	(432)	(2124)
268·800	364	0·774	291·4	72·2	+20·8	6	32	
	364	0·714	289·6	67·0	+18·7	13	36	
	364	0·653	291·4	61·7	+19·0	5	23	
	365	0·607	296·7	57·3	+21·4	0	14	
	368	0·758	121·3	33·1	−18·2	33	165	248 c
	369	0·819	119·1	332·8	−19·2	7	27	
	369	0·845	118·3	330·0	−19·7	0	25	424 c
	369	0·860	118·5	328·4	−20·3	4	11	
	370	0·872	74·6	320·8	+16·7	7	41	
	370	0·904	75·2	316·5	+16·3	7	23	806 c
Sept. 26						(82)	(397)	(1478)
269·907		0·927	245·5					147
	360	0·997	287·0	94·3	+17·4	0	28	447 nf
	364	0·915	291·2	74·1	+22·1	0	8	
	364	0·910	289·7	73·4	+20·7	1	35	
	364	0·932	290·9	72·4	+16·7	10	40	
	364	0·860	287·9	66·9	+18·8	3	18	
	364	0·855	288·9	66·2	+20·5	0	4	
	364	0·815	288·8	61·9	+19·2	0	8	871 f
	366	0·450	198·4	15·9	−18·5	6	32	
	367	0·479	158·7	356·7	−19·8	2	54	
	368	0·592	133·5	340·6	−17·9	31	173	
	369	0·699	126·6	331·0	−19·2	16	84	293 f
	370	0·741	74·2	319·6	+16·3	27	75	249 c
		0·913	80·5					99
		0·958	70·0					486
Sept. 27						(96)	(559)	(2592)
270·808		0·926	289·3					1024
	366	0·553	220·0	17·5	−18·8	4	21	
	366	0·534	215·6	14·6	−19·3	0	9	
	367	0·416	189·0	359·0	−17·4	0	42	
	367	0·435	185·7	358·0	−18·9	0	9	
	367	0·461	183·4	357·1	−20·5	0	16	
	367	0·455	177·9	354·4	−20·2	29	95	
	368	0·475	150·4	341·2	−17·8	55	199	
	368	0·503	146·7	338·5	−18·4	0	15	
	369	0·521	144·5	336·8	−18·7	0	8	
	369	0·531	141·1	334·9	−18·1	0	12	
	369	0·564	138·0	332·0	−18·6	31	129	
	369	0·596	136·5	329·7	−19·5	45	172	
	370	0·534	68·2	324·3	+17·2	12	38	
	370	0·579	69·6	320·9	+17·2	1	11	
	370	0·608	71·1	319·0	+16·6	4	20	
	370	0·603	72·8	318·7	+15·7	9	37	
	370	0·629	73·1	316·9	+15·8	14	53	
		0·921	68·7					515
Sept. 28						(204)	(886)	(1539)

Mean Solar Time.	No. of Group, and Letter for Spot.	Distance from Centre in terms of Sun's Radius.	Position Angle from Sun's Axis.	HELIOGRAPHIC Longitude.	Latitude.	SPOTS. Area of UMBRA for each Spot (and for Day).	Area of WHOLE Spot for each Spot (and for Day).	FACULÆ. Area for each Group (and for Day).
1880. 272d·077		0·960	291·0					290
		0·853	248·0					283
	366	0·659	229·7	10·8	−19·5	0	13	
	366	0·641	227·1	8·5	−20·0	0	15	
	367	0·546	221·1	0·8	−18·0	0	27	
	367	0·550	217·4	359·3	−19·6	15	143	
	367	0·540	212·0	356·4	−20·8	4	29	
	367	0·505	213·4	355·6	−18·5	12	72	
	367	0·511	207·6	353·2	−20·4	83	374	
	368	0·419	186·0	341·3	−17·8	36	162	
	368	0·418	182·6	339·8	−17·9	7	31	
	369	0·435	170·0	334·1	−18·6	0	14	
	369	0·442	167·2	332·7	−18·8	0	11	
	369	0·445	165·2	331·8	−18·7	29	100	
	369	0·478	161·1	329·2	−19·2	34	126	
	370	0·297	52·5	324·5	+16·9	68	291	
	370	0·243	55·6	321·5	+17·9	0	13	
	370	0·366	61·8	319·1	+16·2	8	137	
	370	0·408	65·6	316·1	+13·9	13	89	
		0·789	65·8					
Sept. 29						(309)	(1647)	420 (993)
272·804	366	0·759	237·1	11·5	−19·5	14	71	
	366	0·732	234·1	8·2	−20·3	2	49	
	367	0·642	231·3	0·6	−17·9	29	125	
	367	0·650	229·1	0·4	−19·5	38	126	
	367	0·663	225·1	358·0	−20·9	0	20	
	367	0·606	226·8	356·8	−18·6	17	60	
	367	0·589	220·6	353·2	−20·5	97	427	
	368	0·462	206·9	341·7	−17·8	43	238	
	368	0·438	190·2	333·8	−18·9	0	4	
	369	0·431	186·7	332·1	−18·6	13	76	
	369	0·444	181·8	329·9	−19·6	13	36	
	369	0·444	179·5	328·9	−19·7	4	28	
	370	0·184	13·5	326·5	+16·9	61	278	
	370	0·189	29·7	323·5	+16·0	4	18	
	370	0·214	28·5	323·0	+17·3	7	25	
	370	0·199	35·1	312·3	+15·9	0	6	
	370	0·213	40·5	310·9	+15·8	0	5	
	370	0·215	43·9	319·5	+15·5	7	25	
	370	0·237	46·0	318·9	+16·0	4	19	
	370	0·244	49·6	318·0	+15·6	0	5	
	370	0·260	52·5	316·8	+15·6	2	7	
	370	0·273	51·3	316·3	+16·3	12	37	
Sept.30						(367)	(1685)	49 (49)
274·021	366	0·916	246·1	15·0	−18·8	0	27	487 p
	366	0·900	245·6	12·7	−18·6	0	23	
	366	0·900	244·5	12·4	−19·6	0	12	
	366	0·869	242·3	7·8	−20·1	2	24	
	367	0·824	241·8	2·9	−18·7	4	23	
	367	0·806	241·1	1·0	−18·3	22	143	
	367	0·790	238·9	358·7	−19·5	0	63	
	367	0·773	240·4	357·8	−17·7	12	23	

* Indian photo.   The Areas of Spots and Faculæ are expressed in Millionths of the Sun's visible Hemisphere.

MEASURES of POSITIONS and AREAS of SPOTS and FACULÆ upon the SUN's DISK on PHOTOGRAPHS taken in the

Mean Solar Time	No. of Group, and Letter for Spot	Distance from Centre in terms of Sun's Radius	Position Angle from Sun's Axis	Heliographic Longitude	Latitude	Area of UMBRA for each Spot (and for Day)	Area of WHOLE for each Spot (and for Day)	Area for each Group (and for Day)
1880.								
274d·021	367	0·759	238·2	355·8	−18·7	7	18	
	367	0·763	235·4	355·1	−20·8	4	35	
	367	0·735	234·2	352·4	−20·4	96	467	
	368	0·608	229·8	342·0	−17·3	40	177	
	368	0·578	225·4	338·6	−17·9	0	12	
	369	0·526	217·8	332·8	−18·3	4	23	
	369	0·510	212·7	329·9	−19·0	2	7	
	370	0·300	307·8	327·4	+16·8	49	250	
	370	0·269	310·9	325·2	+16·5	5	11	
	370	0·260	320·2	323·1	+18·0	0	4	
	370	0·254	324·8	321·9	+18·4	0	2	
	370	0·253	331·8	320·3	+19·4	0	6	
	370	0·209	324·7	320·2	+16·3	8	19	
	370	0·211	332·8	318·8	+17·3	0	15	
	370	0·181	332·0	318·1	+15·7	9	19	
	370	0·196	337·5	317·5	+17·0	0	15	
	370	0·175	342·1	316·2	+16·1	6	16	
	371	0·476	57·4	287·8	+20·8	5	9	
	371	0·520	52·0	286·4	+24·5	0	2	
		0·877	111·7					143
		0·944	131·7					80
		0·959	110·0					454
Oct. 1						(275)	(1445)	(1164)
274·896	367	0·903	245·7	1·5	−18·6	0	123	
	367	0·830	230·8	351·3	−20·5	80	435	
	368	0·730	238·6	342·0	−17·4	16	164	
	368	0·608	234·0	338·0	−18·9	1	14	
	369	0·646	231·6	333·6	−18·1	0	6	
	370	0·456	294·6	327·0	+16·8	63	349	
	370	0·397	302·9	322·0	+18·5	0	10	
	370	0·315	304·5	317·1	+16·5	0	35	
Oct. 2						(160)	(1136)	
275·785	367	0·962	248·2	0·3	−18·9	40	346	795 c
	367	0·941	243·9	355·1	−21·9	10	35	
	367	0·919	243·5	351·4	−21·2	125	490	
	368	0·844	243·9	342·3	−17·9	44	228	
	368	0·822	243·4	333·9	−17·5	6	24	
	368	0·811	242·6	338·6	−17·7	0	7	
	368	0·809	240·1	337·6	−19·5	5	9	
	369	0·738	235·1	329·6	−20·0	2	12	1073 p
	370	0·617	288·2	327·2	+16·3	104	477	
	370	0·589	286·0	325·4	+14·7	5	15	
	370	0·589	290·9	324·8	+17·5	9	39	
	370	0·558	292·8	323·3	+18·0	4	28	
	370	0·498	289·7	318·7	+15·3	18	110	
	374	0·920	112·7	226·8	−17·5	0	9	
	374	0·925	110·2	225·4	−15·9	14	73	850 c
Oct. 3*						(386)	(1902)	(2618)
276·768	367	0·982	247·0	352·3	−21·1	103	587	767 e
	368	0·935	248·0	342·3	−18·0	42	170	
	369	0·845	241·9	328·9	−19·6	0	12	
	370	0·764	285·7	326·6	+16·1	80	436	
	374	0·825	114·3	225·6	−15·8	8	34	452 sf
Oct. 4*						(233)	(1239)	(1219)

Mean Solar Time	No. of Group, and Letter for Spot	Distance from Centre in terms of Sun's Radius	Position Angle from Sun's Axis	Heliographic Longitude	Latitude
1880.					
277d·738	368	0·988	250·4	342·1	−18·1
	370	0·884	284·9	326·6	+16·1
	374	0·699	120·8	225·5	−16·0
Oct. 5*					
278·786	370	0·967	285·2	326·4	+16·3
	373	0·536	195·1	159·0	−24·8
	373	0·563	189·6	256·0	−26·0
	374	0·546	132·0	225·4	−15·7
	375	0·906	68·5	184·9	+22·2
	375	0·908	71·6	184·6	+19·3
Oct. 6*					
280·075		0·849	320·5		
	372	0·837	242·7	284·7	−18·7
	373	0·639	216·2	257·7	−25·3
	373	0·632	210·8	254·3	−27·0
	374	0·397	164·1	226·7	−16·2
	374	0·392	159·5	225·0	−15·3
	374	0·364	155·2	224·2	−13·1
	374	0·449	157·7	222·8	−18·3
	374	0·501	163·3	223·7	−22·2
	375	0·766	66·6	184·1	+21·8
	376	0·914	74·3	171·9	+16·9
Oct. 7					
281·051		0·875	320·7		
	372	0·936	246·6	285·6	−19·4
	373	0·757	227·6	258·6	−25·8
	373	0·737	223·0	259·7	−27·5
	374	0·354	200·6	227·6	−13·2
	374	0·373	194·2	225·7	−15·0
	374	0·402	187·5	223·4	−17·2
	375	0·600	60·2	186·1	+22·4
	375	0·619	62·3	184·3	+21·7
	375	0·644	58·4	183·3	+24·6
	377	0·916	62·1	154·1	+28·0
		0·821	151·0		
Oct. 8					
281·832	374	0·439	216·0	225·3	−14·8
	375	0·455	52·3	187·2	+21·8
	375	0·488	52·2	185·3	+22·9
	375	0·491	54·9	184·4	+21·9
	375	0·329	50·4	183·3	+25·1
Oct. 9*					
282·784	373	0·899	236·3	254·0	−26·6
	373	0·764	226·0	235·4	−27·2
	378	0·452	215·8	213·3	−15·6
	378	0·430	209·3	210·0	−16·0
	375	0·316	31·0	187·4	+21·7
	375	0·340	34·8	185·4	+22·1

* Indian photo.   The Areas of Spots and Faculæ are expressed in Millionths of the Sun's visible Hemisphere

MEASURES of POSITIONS and AREAS of SPOTS and FACULÆ upon the SUN's DISK on PHOTOGRAPHS taken in the YEAR 1880—*continued.*

Mean Solar Time	No. of Group and Letter for Spot	Distance from Centre in terms of Sun's Radius	Position Angle from Sun's Axis	Heliographic Longitude	Heliographic Latitude	Area of UMBRA for each Spot (and for Day)	Area of WHOLE Spot (and for Day)	Faculæ — Area for each Group (and for Day)
1880.								
281d·784	375	0·349	38·0	184·1	+21·9	2	6	
	375	0·389	36·2	182·9	+24·1	3	12	
	375	0·408	34·8	182·5	−25·4	2	7	
	375	0·412	38·4	181·1	+24·7	4	15	
Oct. 10*						(28)	(130)	(592)
283·986								970
	373	0·831	270·7	239·9	−26·4	0	5	
	374	0·903	235·8	245·3	−15·3	0	15	
	378	0·790	245·3	229·4	−15·3	6	29	
	378	0·625	235·2	213·7	−15·8	0	16	
	378	0·585	232·7	210·3	−15·5	0	13	
	375	0·289	341·4	187·2	+21·8	0	6	
	375	0·284	344·9	186·1	+21·8	0	3	
	375	0·279	351·7	184·0	+21·9	0	4	
	375	0·374	3·6	180·1	+27·8	5	25	
	375	0·321	8·2	178·7	+24·4			251
		0·879	150·7					264
Oct. 11		0·949	94·2			(11)	(116)	(1485)
284·931	374	0·892	249·6	228·9	−15·2	1	26	916 s f
	378	0·767	243·8	214·5	−15·7	0	5	
	379	0·722	307·9	210·4	+30·8	0	4	125 n p
	379	0·714	307·1	210·0	+29·9	0	5	
	375	0·402	315·7	186·7	+22·3	1	16	1209 s p
Oct. 12	380	0·966	67·6	93·1	+23·1	31	196	(2250)
						(34)	(252)	
286·056	374	0·848	316·8	231·9	−15·8	0	49	303
	374	0·985	252·8	231·9	−15·8	0	41	
	374	0·972	250·9	227·8	−16·9	0	29	
	378	0·896	250·0	224·7	−15·0	0	22	1095 s
	378	0·869	250·4	211·5	−13·8	46	179	
	380	0·881	66·7	92·8	+23·2	0	52	1466 c
Oct. 13	380	0·966	67·9	78·3	+23·0	(46)	(372)	(2864)
286·712	374	0·959	244·6	227·8	−17·6	19	69	490
	378	0·995	251·6	215·0	−15·4	18	45	
	378	0·952	251·7	211·0	−14·0	3	15	547 e
	380	0·928	252·3	92·5	+23·3	53	234	902 f
	380	0·811	65·3	78·9	+21·6	0	20	
	380	0·917	67·9	78·7	+13·5	5	15	260 e
	380	0·919	66·9	76·8	+22·9	0	14	
	380	0·930	67·7	64·1	+21·2	26	128	554 s
Oct. 14*	381	0·986	69·4			(124)	(560)	(2753)
287·697		0·992	253·5	92·9	+23·5	42	183	806
	380	0·674	60·8	91·1	+22·4	2	22	
	380	0·688	63·1	89·1	+19·6	0	7	520 f
	380	0·702	57·7	64·0	+21·1	12	61	
	381	0·918	59·5	56·2	+21·1	0	25	603 c
Oct. 15*	381	0·968	69·6			(56)	(298)	(1929)

Mean Solar Time	No. of Group and Letter for Spot	Distance from Centre in terms of Sun's Radius	Position Angle from Sun's Axis	Heliographic Longitude	Heliographic Latitude	Area of UMBRA for each Spot (and for Day)	Area of WHOLE Spot (and for Day)	Faculæ — Area for each Group (and for Day)
1880.								
288d·749								245
	380	0·919	291·9	109·0	+24·3	3	15	
	380	0·360	25·8	107·8	+24·4	0	4	
	380	0·369	28·4	92·4	+23·9	34	168	
	380	0·523	51·6	92·4	+23·9	3	9	
	380	0·540	56·1	90·0	+22·4	3	7	
	380	0·538	60·3	89·1	+20·3	0	68	
	381	0·822	67·8	64·3	+21·4	16	22	638 s
Oct. 16*	381	0·889	68·9	56·2	+21·3	4	(293)	(883)
						(60)		
289·786		0·964	291·7					617
	380			92·1	+24·1	34	168	
Oct. 17*	381	0·681	64·1	64·0	+21·4	21	81	(617)
						(55)	(249)	
291·056		0·909	325·0					439
	380	0·381	34·5	93·2	+22·8	0	3	
	380	0·332	350·4	91·8	+24·4	27	133	
	381	0·480	53·4	63·9	+21·5	18	72	
	381	0·722	61·6	44·5	+24·0	0	11	97 f
Oct. 18		0·959	140·5			(45)	(219)	561
								(1097)
291·784	380	0·388	327·4	91·9	+24·2	43	145	
	380	0·303	327·2	88·7	+20·0	3	19	
	380	0·287	326·4	88·3	+19·0	14	34	
	380	0·273	327·0	87·7	+18·4	4	18	
	380	0·303	313·9	91·9	+17·2	25	78	
	380	0·315	312·3	91·9	+18·9	4	16	
	381	0·365	40·3	64·1	+21·2	27	75	
	382a	0·993	109·8	358·0	−18·9	0	161	
Oct. 19*	382a	0·995	111·9	357·2	−21·1	19	302	793 s p
						(139)	(848)	(793)
293·784	380	0·686	291·5	94·4	+18·4	62	310	
	380	0·623	296·0	88·7	+20·0	13	37	
	380	0·609	297·4	87·4	+20·5	2	14	
	380	0·596	296·6	86·6	+19·8	16	98	
	380	0·671	301·0	91·2	+24·2	31	125	
	382b	0·385	307·4	71·0	+18·4	11	45	
	382b	0·355	313·3	68·1	+19·0	8	42	
	381	0·331	325·8	63·7	+21·0	24	75	
	381	0·324	330·0	62·3	+21·4	2	10	
	382a	0·857	116·2	358·0	−19·1	25	156	
Oct. 21*	382a	0·863	119·3	358·2	−22·0	37	240	803 f
						(231)	(1152)	(803)
294·790	380	0·834	288·4	95·1	+18·1	55	245	
	380	0·766	293·0	87·7	+20·7	0	9	
	380	0·804	296·5	90·9	+20·8	20	66	
	380	0·745	292·5	85·9	+20·1	11	46	
	382b	0·571	295·1	71·9	+18·3	28	130	
	382b	0·559	300·2	70·0	+20·7	0	13	
	382b	0·525	296·9	68·4	+18·1	1	7	

* Indian photo.   The Areas of Spots and Faculæ are expressed in Millionths of the Sun's visible Hemisphere.

MEASURES of POSITIONS and AREAS of SPOTS and FACULÆ upon the SUN's DISK on PHOTOGRAPHS taken in the YEAR 1880—*continued.*

Mean Solar Time.	No. of Group, and Letter for Spot.	Distance from Centre in terms of Sun's Radius.	Position Angle from Sun's Axis.	Longitude.	Latitude.	Area of UMBRA for each Spot (and for Day).	Area of WHOLE Spot (and for Day).	Area for each Group (and for Day).
1880.								
294d·790	382b	0·521	300·0	67·5	+19·6	18	74	
	381	0·480	305·6	63·6	+20·8	17	58	
	382a	0·729	122·4	358·5	−19·1	23	63	
	382a	0·758	123·1	356·4	−20·7	0	14	
	382a	0·751	125·9	358·1	−21·2	46	208	
	382a	0·772	124·6	355·8	−22·2	0	12	
Oct. 22*	382c	0·947	74·7	327·3	+16·1	0	40	351 n
						(219)	(985)	(351)
295·679	380	0·927	287·2	93·5	+17·8	50	296	
	380	0·901	294·9	91·0	+24·5	5	30	
	380	0·876	290·5	90·6	+20·3	0	8	
	380	0·855	290·9	85·5	+20·4	2	11	996 nf
	382b	0·722	290·7	72·5	+18·3	50	227	
	382b	0·687	291·6	69·6	+18·4	0	10	
	382b	0·682	293·2	68·7	+19·3	11	39	
	382b	0·661	294·3	66·9	+19·7	16	79	
	381	0·628	297·9	63·6	+21·1	11	25	
	381	0·612	298·7	62·3	+21·2	6	16	
	381	0·594	300·8	60·5	+21·8	0	5	
	382a	0·660	136·5	0·9	−21·2	2	11	
	382a	0·607	131·8	358·5	−19·5	15	72	
	382a	0·638	134·9	357·9	−22·4	71	360	
	382a	0·664	137·4	357·7	−24·9	2	6	
	382c	0·865	74·6	327·5	+15·8	4	17	260 nf / 542
Oct. 23*		0·865	119·3					(1798)
						(245)	(1212)	
296·783	380	0·991	287·1	96·1	+17·6	43	151	1781 nf
	382b	0·866	288·0	72·6	+18·0	69	336	
	382b	0·840	288·0	69·6	+17·8	0	20	
	382b	0·827	288·7	68·2	+18·2	7	21	
	382b	0·834	290·5	68·5	+19·7	4	16	
	382b	0·814	291·0	66·5	+19·9	12	39	
	381	0·792	293·2	63·9	+21·2	6	12	
	381	0·770	293·2	62·7	+21·1	4	8	708 e
	382a	0·473	155·3	0·7	−20·6	2	9	
	382a	0·491	155·1	0·0	−21·5	5	17	
	382a	0·489	152·2	358·7	−20·7	3	5	
	382a	0·473	150·5	358·6	−19·5	26	72	
	382a	0·519	151·4	357·8	−22·5	59	320	
	382c	0·717	72·3	327·7	+16·0	6	17	
Oct. 24*						(246)	(1043)	(2489)
297·886	382b	0·958	287·1	72·1	+17·7	78	263	
	381	0·906	291·5	63·0	+21·5	0	14	1090 f
	382a	0·437	182·4	359·4	−21·0	2	7	
	382a	0·413	180·5	358·5	−19·5	14	45	
	382a	0·433	179·7	358·2	−20·8	8	30	
	382a	0·460	178·8	357·7	−22·5	37	215	
Oct. 25*								
298·784	382b	0·995	287·8	71·4	+18·2	47	199	1062 nf
	382a	0·457	206·0	358·6	−19·5	9	31	
	382a	0·488	203·1	358·3	−22·0	32	125	
Oct. 26*						(88)	(355)	(1062)

Mean Solar Time.	No. of Group, and Letter for Spot.	Distance from Centre in terms of Sun's Radius.	Position Angle from Sun's Axis.	Longitude.	Latitude.	Area of UMBRA for each Spot (and for Day).	Area of WHOLE Spot (and for Day).	Area for each Group (and for Day).
1880.								
299d·805	382a	0·614	221·6	359·1	−23·2	3	12	
	382a	0·589	223·8	358·7	−20·9	40	168	
	382a	0·575	225·9	358·7	−19·3	5	19	
	382d	0·447	227·4	352·6	−13·2	8	26	
	382a	0·400	222·8	349·0	−12·5	8	24	
	382e	0·518	75·0	302·5	+11·6	2	11	
Oct. 27*		0·948	64·3					98
						(66)	(260)	(98)
300·785	382a	0·734	236·7	0·7	−20·4	30	88	
	382a	0·714	255·7	358·8	−20·1	4	24	
	382a	0·708	234·6	357·9	−20·6	11	29	
	382e	0·322	66·6	302·5	+11·6	5	25	
Oct. 28*						(50)	(166)	
301·785	382a	0·857	243·8	1·4	−19·7	15	42	219 c
	382f	0·524	297·7	335·9	+17·9	4	20	
	382c	0·107	301·4	334·0	+18·6	2	9	
	382c	0·107		306·1	+10·5	6	19	
	382e	0·126	14·8	304·9	+11·4	2	8	
	382e	0·142	28·7	302·8	+11·6	10	29	
Oct. 29*						(39)	(127)	(219)
303·063	383	0·899	244·3	235·8	−22·6	0	40	455
		0·862	119·3					603 nf
		0·961	108·2					277
Oct. 30						(0)	(40)	(1335)
303·793		0·958	246·0					1269
	383	0·751	117·1	236·0	−17·0	9	21	
	383	0·778	124·2	236·2	−22·9	3	12	
	383	0·785	123·4	235·2	−22·6	18	35	401 f
		0·926	110·0					721
Oct. 31*						(32)	(68)	(2391)
304·783		0·932	243·6					514
	383a	0·864	292·7	326·1	+21·6	0	31	
	384	0·608	290·2	303·4	+15·4	4	18	
	384	0·665	292·4	302·9	+16·6	2	10	
	384	0·592	290·9	302·2	+15·6	14	25	
	384	0·596	293·0	302·1	+16·8	2	9	
	384	0·582	294·9	300·8	+17·5	6	21	
	384	0·560	295·4	299·2	+17·4	0	4	
	384	0·545	292·6	298·7	+15·6	2	9	
	385	0·602	125·9	236·7	−17·1	2	16	
	383	0·654	132·7	236·0	−22·8	2	6	
	383	0·662	132·0	235·2	−22·7	6	19	
	383	0·676	131·0	233·8	−22·8	2	15	
	383b	0·756	105·8	220·0	−9·1	2	7	
	386	0·936	112·2	201·0	−19·1	13	49	
		0·911	115·3					643
Nov. 1*						(57)	(239)	(1157)

* Indian photo.   The Areas of Spots and Faculæ are expressed in Millionths of the Sun's visible Hemisphere.

MEASURES of POSITIONS and AREAS of SPOTS and FACULÆ upon the SUN'S DISK on PHOTOGRAPHS taken in the YEAR 1880—*continued.*

Mean Solar Time.	No. of Group, and Letter for Spot.	Distance from Centre in terms of Sun's Radius.	Position Angle from Sun's Axis.	Heliographic Longitude.	Latitude.	Spots. Area of UMBRA for each Spot (and for Day).	Area of WHOLE Spot (and for Day).	Faculæ. Area for each Group (and for Day).
1880. 305d.978		0.991	250.5					252
		0.953	288.2					372
	384	0.812	285.9	305.3	+15.2	10	87	
	384	0.787	287.1	302.8	+15.9	4	17	
	384	0.788	290.3	302.5	+18.4	15	34	
	384	0.778	288.6	301.7	+16.9	0	21	
	384	0.737	289.7	297.9	+17.2	6	20	323 c
	383	0.518	148.8	234.7	-22.4	0	14	
	385	0.419	147.3	237.9	-16.7	0	12	
	385	0.399	143.3	237.3	-14.7	0	13	
	385	0.427	145.6	237.0	-16.7	11	47	
	385	0.427	141.4	235.5	-15.6	0	11	
	385	0.448	140.7	234.4	-16.4	0	18	
	385	0.459	139.9	233.6	-16.7	0	4	
	386	0.819	115.9	200.8	-18.4	4	21	584 c
	387	0.930	107.7	185.4	-14.8	0	9	156 sf
		0.927	64.7					205
Nov. 2						(50)	(328)	(1892)
306.969		0.819	296.0					96
	384	0.920	284.5	305.4	+14.8	12	45	
	384	0.905	286.3	303.1	+16.6	0	4	580 c
	384	0.883	287.9	300.1	+17.6	0	10	
	384	0.818	284.5	293.0	+14.1	0	9	
	385	0.360	176.1	237.0	-17.1	16	40	
	385	0.384	170.2	234.5	-18.2	0	5	
	385	0.371	164.9	232.7	-17.0	7	52	215 f
	386	0.698	122.2	200.0	-18.7	5	24	68 f
	387	0.835	111.8	185.0	-15.7	3	10	176
		0.846	62.3					(1135)
Nov. 3						(43)	(199)	
308.075		0.961	288.0					526
		0.904	247.0					86
		0.770	235.5					224
	385	0.421	212.1	237.3	-17.2	8	17	
	385	0.404	210.3	236.1	-16.7	0	5	
	385	0.415	208.9	235.9	-17.5	0	7	
	385	0.387	202.1	232.6	-17.2	8	20	
	385	0.390	199.7	231.7	-17.7	4	16	
	388	0.434	40.3	206.7	+22.4	2	8	
	386	0.540	134.4	199.9	-18.7	7	21	117 sf
	387	0.686	117.9	185.0	-15.8	0	5	
		0.714	58.2					213
		0.939	67.0					342
Nov. 4						(29)	(99)	(1508)
308.750	385	0.504	225.9	237.1	-17.1	4	20	
	385	0.457	219.5	232.5	-17.1	11	28	
	385	0.454	217.3	231.6	-17.5	4	14	
	386	0.451	147.7	200.2	-18.7	7	19	
		0.925	67.2					408
Nov. 5*						(22)	(91)	(408)

Mean Solar Time.	No. of Group, and Letter for Spot.	Distance from Centre in terms of Sun's Radius.	Position Angle from Sun's Axis.	Heliographic Longitude.	Latitude.	Spots. Area of UMBRA for each Spot (and for Day).	Area of WHOLE Spot (and for Day).	Faculæ. Area for each Group (and for Day).
1880. 311d.904		0.915	251.2					323
	389	0.972	138.5	245.4	-29.6	0	14	
	389	0.968	239.2	244.6	-28.7	0	16	544 nf
		0.953	69.5					492
Nov. 8						(0)	(30)	(1359)
312.904		0.957	245.3					528
		0.875	249.3					160
		0.785	239.2					208
	390	0.483	124.0	136.1	-12.7	0	11	
	391	0.694	80.2	116.6	+9.1	2	11	768
		0.889	69.3					246
Nov. 9		0.984	111.2			(2)	(22)	(1910)
316.783		0.941	295.3					274
	392	0.812	66.1	56.6	+20.9	3	24	
	392	0.844	66.3	53.2	+21.4	12	40	331 c
Nov.13*						(15)	(64)	(605)
317.844	392	0.657	59.2	57.7	+21.8	41	114	
	392	0.683	59.7	55.6	+22.2	21	64	366 c
	392	0.711	61.3	52.9	+21.9	7	42	
	392a	0.908	108.9	32.0	-15.8	0	7	367
Nov.14*						(69)	(227)	(733)
318.800		0.922	296.7					219
	392	0.523	50.4	56.9	+21.7	152	491	
	392	0.567	53.5	53.2	+21.9	19	57	
		0.903	116.7					396
Nov.15*						(171)	(548)	(615)
319.948		0.941	304.5					179
	392	0.379	16.3	57.0	+22.2	68	419	
		0.941	114.5					143
Nov. 16						(68)	(419)	(322)
320.795	392	0.357	358.4	56.8	+21.9	88	356	
	394	0.364	135.7	41.2	-12.8	0	8	
	394	0.368	132.1	40.1	-12.1	7	25	
	394	0.411	132.5	37.8	-13.8	26	53	
Nov.17*						(121)	(442)	
321.576	392	0.399	331.3	56.7	+22.7	22	201	
	392	0.371	335.9	54.2	+22.0	0	82	
	393	0.278	220.1	55.2	-9.9	0	19	
	393	0.232	110.2	51.6	-9.2	0	47	
	394	0.264	166.6	41.2	-12.5	0	19	
	394	0.311	156.1	37.4	-14.2	0	39	
	394	0.310	153.3	36.6	-13.7	16	67	
Nov.18†						(38)	(474)	

* Indian photo.   The Areas of Spots and Faculæ are expressed in Millionths of the Sun's visible Hemisphere.   † Melbourne photo.

MEASURES of POSITIONS and AREAS of SPOTS and FACULÆ upon the SUN'S DISK on PHOTOGRAPHS taken in the YEAR 1880—*continued.*

Mean Solar Time.	No. of Group, and Letter for Spot.	Distance from Centre in terms of Sun's Radius.	Position Angle from Sun's Axis.	Heliographic Longitude.	Heliographic Latitude.	Area of UMBRA for each Spot (and for Day).	Area of WHOLE Spot for each Spot (and for Day).	Faculæ. Area for each Group (and for Day).
1880.			°	°	°			
323ᵈ·045		0·962	265·2					125
		0·920	290·2					263
		0·854	242·3	—				136
	392	0·579	307·2	36·3	+22·2	39	230	
	393	0·554	247·6	57·8	—10·4	40	158	
	393	0·516	249·5	55·7	— 8·6	0	8	
	393	0·492	248·3	53·9	— 8·7	0	7	
	393	0·467	245·3	51·9	— 9·4	8	142	
	394	0·379	229·2	43·5	—12·3	6	40	
	394	0·329	216·5	38·1	—13·3	17	96	
	394	0·330	210·1	36·3	—14·5	0	17	
	395	0·729	62·9	342·7	+20·8	0	6	
		0·781	63·5					78
		0·843	54·7					90
		0·854	113·2					167
		0·916	68·3					110
Nov. 19						(110)	(704)	(969)
324·058		0·941	245·5					201
		0·903	292·5					239
	392	0·723	299·2	55·9	+22·0	23	238	250 c
	393	0·760	261·5	62·0	— 5·2	0	8	
	393	0·737	254·1	58·2	—10·1	45	218	
	393	0·650	253·1	52·1	— 9·5	27	185	449 c
	394	0·561	244·2	44·2	—12·5	12	28	
	394	0·483	237·8	37·9	—13·1	33	105	
Nov. 20						(140)	(782)	(1140)
325·884	392	0·924	293·2	55·2	+21·0	61	242	680 c
	393	0·939	259·3	58·2	— 9·5	58	179	
	393	0·889	258·7	50·9	— 9·2	29	95	1124 c
Nov. 22*						(148)	(516)	(1804)
326·809	392	0·984	292·0	56·2	+21·9	60	111	171 f
	393	0·989	260·2	57·8	— 9·5	41	205	
	393	0·968	260·3	51·7	— 9·0	24	73	1664 c
Nov. 23*						(125)	(389)	(1835)
327·802		0·981	255·5					867
	396	0·859	110·3	266·9	—16·5	3	15	
	396	0·890	110·2	263·0	—17·2	7	31	
	397	0·933	66·7	256·3	+22·2	60	248	
	397	0·948	68·8	253·4	+20·5	58	266	
	397	0·952	67·0	252·8	+22·3	0	60	
	397	0·958	65·8	251·7	+23·5	54	230	1727 c
	397	0·970	67·7	248·6	+22·0	75	270	
Nov. 24*								
329·015	395	0·623	302·9	341·7	+20·8	6	27	
	396	0·696	117·4	267·6	—17·6	8	38	
	396	0·739	116·0	263·7	—17·9	11	34	
	396	0·742	113·8	263·0	—16·4	10	25	
	396	0·754	114·4	262·1	—17·2	6	15	228 s f

Mean Solar Time.	No. of Group, and Letter for Spot.	Distance from Centre in terms of Sun's Radius.	Position Angle from Sun's Axis.	Heliographic Longitude.	Heliographic Latitude.	Area of UMBRA for each Spot (and for Day).	Area of WHOLE Spot for each Spot (and for Day).	Faculæ. Area for each Group (and for Day).
1880.			°	°	°			
329ᵈ·015	397	0·817	64·8	255·6	+21·2	122	700	
	397	0·877	65·1	248·8	+22·3	76	538	732 c
	398	0·970	115·6	234·2	—24·4	0	20	537 s j
		0·989	124·8					187
Nov. 25						(239)	(1397)	(1684)
329·682	396	0·597	123·4	267·5	—18·1	14	35	
	396	0·641	122·3	264·2	—19·0	2	10	
	396	0·656	119·5	262·3	—17·8	17	68	
	397	0·715	65·2	255·9	+18·3	0	38	
	397	0·733	65·4	254·5	+18·5	0	16	
	397	0·732	61·2	255·6	+21·4	126	754	
	397	0·751	65·1	253·0	+19·2	0	7	
	397	0·786	61·6	250·7	+22·5	25	95	
	397	0·788	62·7	250·2	+21·9	33	164	
	397	0·798	64·2	249·8	+21·1	8	38	
	397	0·808	63·0	248·2	+22·2	8	29	
	397	0·808	63·6	248·0	+21·7	11	26	
	397	0·798	66·7	248·3	+19·2	3	21	
	397	0·823	64·7	246·2	+21·3	49	228	
	397	0·819	66·8	246·2	+19·5	3	13	
	397	0·831	66·5	244·9	+20·1	3	20	709 c
	398	0·922	117·3	235·0	—24·5	4	40	924 c
Nov. 26*						(306)	(1602)	(1633)
331·052	395	0·907	293·2	344·3	+21·4	0	9	428 n
	399	0·866	294·2	338·8	+21·3	0	6	
	399	0·864	243·6	337·4	—21·9	0	15	311 c
	396	0·369	146·2	268·7	—16·8	0	2	
	396	0·388	145·1	267·6	—17·5	3	16	
	396	0·440	136·0	262·4	—17·4	0	2	
	397	0·455	43·2	261·7	+20·3	0	10	
	397	0·489	52·7	257·0	+18·1	4	19	
	397	0·519	49·3	256·2	+20·6	103	746	
	397	0·505	53·7	255·8	+18·3	0	12	
	397	0·514	55·1	254·8	+18·0	3	25	
	397	0·556	53·9	252·6	+19·9	0	6	
	397	0·556	56·1	252·0	+18·9	0	10	
	397	0·602	53·5	249·7	+21·8	11	37	
	397	0·615	55·2	248·3	+21·4	0	17	
	397	0·623	52·7	248·6	+23·0	16	39	
	397	0·611	57·9	247·4	+21·5	0	2	
	397	0·635	54·0	247·3	+21·7	5	28	
	397	0·646	56·4	245·8	+21·7	37	237	
	397	0·668	57·5	243·8	+21·8	3	27	
	398	0·785	123·1	234·9	—24·6	7	16	312 s j
Nov. 27						(192)	(1281)	(1051)
331·784	397	0·945	292·8					566
	397	0·390	33·1	258·3	+19·8	14	59	
	397	0·406	34·6	257·1	+20·3	152	665	
	397	0·423	36·6	255·7	+20·6	31	95	
	397	0·435	38·7	254·5	+20·6	5	29	
	397	0·500	44·4	249·3	+21·7	0	22	

* Indian photo.   The Areas of Spots and Faculæ are expressed in Millionths of the Sun's visible Hemisphere.

MEASURES of POSITIONS and AREAS of SPOTS and FACULÆ upon the Sun's DISK on PHOTOGRAPHS taken in the YEAR 1880—*continued.*

Mean Solar Time.	No. of Group, and Letter for Spot.	Distance from Centre in terms of Sun's Radius.	Position Angle from Sun's Axis.	Heliographic Longitude.	Heliographic Latitude.	Area of UMBRA for each Spot (and for Day).	Area of WHOLE Spot (and for Day).	Area for each Group (and for Day).
1880.								
331ᵈ·784	397	0·519	43·6	246·5	+22·8	20	60	
	397	0·534	45·7	247·0	+22·6	0	55	
	397	0·541	48·4	245·6	+21·8	65	185	
	397	0·557	51·6	243·5	+21·0	6	21	
	398	0·674	128·8	236·4	−24·1	2	15	
	398	0·695	128·1	234·5	−24·6	0	6	
Nov. 28*						(295)	(1212)	(566)
332·903		0·784	295·5					88
		0·780	286·3					52
	397	0·299	355·7	257·9	+18·1	2	23	
	397	0·280	357·3	257·4	+17·0	0	5	
	397	0·333	0·1	256·5	+20·2	161	688	
	397	0·324	9·1	253·5	+19·4	0	6	
	397	0·344	11·5	252·4	+20·4	3	13	
	397	0·380	18·1	249·3	+21·9	0	15	
	397	0·393	21·0	247·9	+22·2	18	57	
	397	0·406	22·6	246·9	+22·7	6	29	
	397	0·408	26·2	245·5	+22·2	43	146	
	397	0·396	32·1	243·7	+20·3	0	7	
	398	0·393	143·3	236·2	−24·4	5	8	
		0·744	62·7					58
		0·814	117·8					45
		0·874	114·5					41
		0·898	132·0					119
Nov. 29						(238)	(997)	(403)
333·988		0·909	285·2					53
		0·901	292·7					153
	397	0·406	326·1	256·2	+20·3	100	614	
	397	0·358	318·5	253·6	+18·3	1	11	
	397	0·342	331·4	252·2	+18·1	0	6	
	397	0·350	345·2	247·8	+20·4	0	6	
	397	0·377	347·9	247·2	+22·2	8	30	
	397	0·369	332·2	245·4	+22·0	14	43	
	397	0·335	334·6	244·2	+20·1	0	6	
	397	0·366	356·2	243·8	+22·0	5	10	
	397	0·421	0·0	242·3	+25·5	1	12	
		0·927	59·3					396
Nov. 30						(129)	(738)	(602)
334·685	397	0·494	311·5	256·2	+19·6	99	600	
	397	0·471	319·3	252·4	+21·4	0	18	
	397	0·428	327·3	247·5	+21·6	9	52	
	397	0·409	330·2	245·8	+21·3	7	34	
Dec. 1*						(115)	(704)	
336·685	397	0·792	295·6	256·1	+20·1	79	446	264 c
	400	0·942	104·3	137·3	−13·3	82	401	
	400	0·985	105·3	127·3	−15·1	0	70	1201 c
	400	0·991	107·9	125·2	−18·7	0	100	
Dec. 3*						(161)	(1017)	(1465)

Mean Solar Time.	No. of Group, and Letter for Spot.	Distance from Centre in terms of Sun's Radius.	Position Angle from Sun's Axis.	Heliographic Longitude.	Heliographic Latitude.	Area of UMBRA for each Spot (and for Day).	Area of WHOLE Spot (and for Day).	Area for each Group (and for Day).
1880.								
337ᵈ·788		0·960	251·5					288
	400	0·954	294·4	267·2	+23·2	0	15	
	397	0·905	292·0	255·1	+19·8	83	504	1670 c
	400	0·801	107·3	140·5	−13·7	3	26	
	400	0·840	105·9	136·2	−13·2	57	433	
	400	0·864	106·3	133·7	−13·9	0	21	
	400	0·870	104·6	133·0	−12·5	3	31	
	400	0·885	108·2	131·5	−15·9	0	3	
	400	0·908	108·4	128·5	−16·5	4	15	
	400	0·924	108·1	126·0	−16·5	24	68	
	400	0·943	109·2	123·0	−18·0	0	9	
Dec. 4*	400	0·944	107·1	122·6	−16·0	18	69	1662 c
						(192)	(1194)	(3620)
339·687	400b	0·340	27·1	157·9	+17·5	0	54	
	400b	0·353	34·3	153·7	+18·8	24	84	
	400	0·559	114·3	135·8	−13·3	80	409	
	400	0·611	113·0	132·0	−13·8	0	18	
	400	0·629	113·0	130·7	−14·3	0	14	
	400	0·664	114·0	128·3	−15·7	0	39	
	400	0·698	111·0	125·8	−14·3	28	64	
	400	0·747	114·8	121·8	−18·3	12	35	630 c
		0·950	128·4					537
Dec. 6*						(144)	(717)	(1167)
340·686	400c	0·861	291·3	211·4	+18·1	9	30	
	400c	0·843	293·9	208·8	+19·9	11	40	757 c
	400b	0·310	338·4	160·8	+16·5	0	16	
	400b	0·301	344·1	158·9	+16·6	11	26	
	400b	0·301	348·8	157·5	+17·0	5	21	
	400b	0·328	0·4	153·9	+18·9	16	62	
	400b	0·363	1·6	153·4	+21·0	0	5	
	400b	0·389	1·5	153·4	+22·0	0	3	
	400	0·384	128·0	135·9	−13·8	84	552	
	400	0·519	118·9	126·1	−14·7	23	54	
	400d	0·702	114·5	112·3	−16·8	2	17	
		0·891	131·8					658
Dec. 7*						(161)	(826)	(1415)
342·784	400	0·290	216·0	136·5	−14·0	87	595	
		0·921	100·2					687
Dec. 9*						(87)	(595)	(687)
343·979		0·926	297·3					310
		0·745	297·2					255
		0·767	234·5					88
	400	0·488	242·0	136·9	−13·8	85	517	k
	400	0·398	217·2	128·3	−16·2	0	8	
	400	0·367	227·9	127·0	−14·8	4	6	
		0·855	104·8					342
		0·925	65·0					562
Dec. 10						(89)	(531)	(1901)

* Indian photo.   The Areas of Spots and Faculæ are expressed in Millionths of the Sun's visible Hemisphere.

MEASURES of POSITIONS and AREAS of SPOTS and FACULÆ upon the SUN'S DISK on PHOTOGRAPHS taken in the YEAR 1880—*continued.*

Mean Solar Time	No. of Group, and Letter for Spot	Distance from Centre in terms of Sun's Radius	Position Angle from Sun's Axis	Longi- tude	Lati- tude	Area of UMBRA for each Spot (and for Day)	Area of WHOLE for each Spot (and for Day)	Area for each Group (and for Day)
1880. 344d·965		0·798	299·5					385
	400	0·693	251·3	140·1	−12·7	0	7	
	400	0·673	251·2	138·3	−13·1	47	190	
	400	0·652	254·3	137·2	−16·7	0	29	292 c
	400	0·641	249·6	135·7	−13·5	73	280	
	400	0·630	252·5	135·3	−11·5	0	26	
		0·853						298
Dec. 11						(120)	(532)	(885)
345·775		0·936	291·3					670
	400	0·817	247·3	139·6	−18·9	3	8	
	400	0·793	253·6	138·3	−13·5	30	126	
	400	0·780	253·1	137·0	−13·6	7	61	
	400	0·767	254·8	136·0	−12·1	5	50	
	400	0·763	252·2	135·3	−14·0	43	202	
Dec. 12*						(88)	(447)	(670)
346·824		0·936	298·5					238
	400	0·911	255·2	138·1	−13·8	37	141	
	400	0·899	256·0	136·3	−12·4	10	59	1194
	400	0·897	256·0	136·3	−13·0	7	48	
	400	0·888	254·2	134·9	−14·4	57	233	
Dec. 13*						(111)	(481)	(1432)
349·734		0·939	291·8					710
		0·899	227·3					454
	402	0·948	75·0	324·4	+13·8	29	86	936 f
Dec. 16*						(29)	(86)	(2100)
350·774		0·953	231·3					454
	402	0·856	73·2	323·8	+13·6	24	68	480 f
	401	0·899	65·2	320·1	+21·4	7	28	
	401	0·915	64·9	317·9	+22·2	11	49	1020 c
	401	0·929	61·6	316·1	+24·7	8	33	
Dec. 17*						(50)	(178)	(1954)
351·779	402	0·715	68·7	324·8	+13·8	22	44	
	401	0·787	59·4	320·9	+22·5	5	22	
	401	0·795	59·1	320·3	+23·0	10	30	
	401	0·832	58·8	316·4	+24·5	6	19	23 g c
	401	0·842	57·9	315·9	+25·5	0	9	
	401	0·847	59·1	314·9	+24·8	3	14	
		0·916	73·5					534
Dec. 18*						(46)	(138)	(773)
352·682	402	0·564	61·9	325·3	+13·8	19	75	
	401	0·670	50·8	321·6	+13·6	19	73	
	401	0·731	52·1	316·5	+25·3	2	11	
Dec. 19*						(40)	(159)	
353·687	402	0·391	47·6	325·5	+13·5	18	80	
	401	0·541	37·1	321·9	+23·8	18	88	
Dec. 20*						(36)	(168)	

Mean Solar Time	No. of Group, and Letter for Spot	Distance from Centre in terms of Sun's Radius	Position Angle from Sun's Axis	Longi- tude	Lati- tude	Area of UMBRA for each Spot (and for Day)	Area of WHOLE for each Spot (and for Day)	Area for each Group (and for Day)
1880. 354d·891	402	0·269	358·8	327·1	+13·6	24	76	
	401	0·448	8·3	322·7	+24·2	16	28	
	403	0·972	68·3	252·2	+20·6	64	277	1058 c
Dec. 21*						(104)	(381)	(1058)
355·784	402	0·337	323·6	326·9	+13·6	16	34	
	401	0·458	346·1	322·0	+24·2	9	24	
	403	0·923	66·4	250·5	+20·8	75	377	1596 f
Dec. 22*						(100)	(435)	(1596)
356·911		0·802	302·3					115
	401	0·561	321·8	322·6	+24·0	4	6	
	401	0·552	324·7	320·8	+24·6	2	16	
	401	0·537	329·2	318·0	+25·2	1	12	
	402	0·526	300·7	327·9	+13·5	4	20	
	402	0·468	306·7	322·7	+14·1	6	11	
	402	0·464	308·1	322·1	+14·5	0	4	
	402	0·420	308·9	319·5	+13·1	0	3	
	402	0·422	310·9	319·1	+13·9	3	9	
	403	0·804	61·8	250·9	+20·8	69	341	293 f
		0·881	114·8					264
Dec. 23						(89)	(422)	(672)
357·939		0·819	238·7					225
	402	0·606	292·3	328·1	+13·5	14	26	
	402	0·653	295·2	324·2	+14·2	5	11	
	402	0·631	295·1	322·6	+13·5	2	8	
	402	0·608	295·1	321·0	+12·9	5	9	
	402	0·585	297·4	319·0	+13·6	12	65	
	401	0·679	311·2	320·8	+24·5	4	21	307 f
	403	0·673	55·1	250·7	+20·6	79	335	240
		0·776	117·6					
Dec. 24						(121)	(475)	(772)
359·664 Dec. 26*	403	0·456	29·4	250·2	+20·8	42 (42)	225 (225)	
360·779		0·951	284·4					1082
	403	0·400	357·7	250·3	+20·8	62	337	
	406	0·874	123·4	192·0	−30·1	57	245	
	406	0·920	121·8	184·8	−30·1	0	20	1338 c
	406	0·932	121·6	182·7	−30·3	4	50	
Dec. 27*						(123)	(653)	(2420)
361·772	403	0·454	332·9	249·0	+21·0	42	204	
	406	0·771	129·2	192·2	−31·1	24	92	
		0·838	125·8					662
Dec.28†						(66)	(296)	(662)
362·743	403	0·569	313·0	249·7	+20·1	56	268	
	406	0·664	135·6	190·9	−30·7	39	103	
Dec. 29*						(95)	(366)	

* Indian photo.   The Areas of Spots and Faculæ are expressed in Millionths of the Sun's visible Hemisphere.   † Melbourne photo.

MEASURES of POSITIONS and AREAS of SPOTS and FACULÆ upon the SUN's DISK on PHOTOGRAPHS taken in the YEAR 1880—*continued.*

Mean Solar Time.	No. of Group and Letter for Spot.	Distance from Centre in terms of Sun's Radius.	Position Angle from Sun's Axis.	HELIOGRAPHIC Longitude.	HELIOGRAPHIC Latitude.	SPOTS Area of UMBRA for each Spot (and for Day).	SPOTS Area of WHOLE for each Spot (and for Day).	FACULÆ Area for each Group (and for Day).
1880.		°	°	°	°			
163ᵈ·789	403	0·711	302·3	249·4	+19·9	42	252	
	405	0·253	171·6	207·6	−17·5	2	18	
	405	0·270	171·5	207·4	−18·5	3	16	
	405	0·289	162·3	204·5	−19·0	6	23	
	406	0·550	149·4	190·7	−31·1	46	185	
		0·945	100·8			(99)	(560)	702 (702)
Dec. 30								
365·078	404	0·378	228·7	210·0	−17·5	47	139	
	404	0·374	224·9	208·8	−18·4	0	9	
	405	0·342	219·8	206·0	−18·3	0	8	
	405	0·342	214·4	204·5	−19·5	46	123	
	406	0·479	175·6	190·2	−31·6	30	153	
	406	0·490	171·2	187·7	−32·0	0	12	
Dec. 31						(123)	(444)	
365·993	403	0·943	293·6	248·1	+20·9	34	214	
	404	0·355	243·1	211·8	−17·4	27	116	
	405	0·474	232·9	204·2	−19·6	20	75	
1881.	406	0·495	196·2	188·0	−31·5	35	71	
Jan. 1						(116)	(476)	
1·709	404	0·833	251·7	213·6	−17·1	56	244	
	404	0·789	250·4	209·0	−17·5	5	20	1161 e
	405	0·771	247·4	206·9	−19·5	0	34	
	405	0·741	247·3	204·1	−19·0	44	160	
	406	0·646	212·7	188·7	−31·3	40	192	
	407	0·928	61·0	94·4	+25·2	115	445	1972 f
Jan. 3*						(260)	(1095)	(3133)
2·679	404	0·930	253·0	213·6	−17·1	46	262	
	405	0·873	249·4	205·3	−19·7	0	60	1515 c
	405	0·855	249·8	203·2	−19·1	21	80	
	406	0·759	231·4	188·8	−30·8	52	165	
	407	0·848	57·2	93·5	+25·1	110	550	2459 f
Jan. 4*						(229)	(1117)	(3974)
4·046	405	0·969	251·1	203·2	−19·2	0	54	139 s p
	406	0·901	237·2	189·1	−31·0	23	88	405 s f
	407	0·702	47·3	92·7	+25·3	93	391	750 s f
Jan. 5						(116)	(533)	(1294)
4·929	406	0·957	239·0	188·0	−30·7	14	73	445 f
	407	0·603	36·1	92·6	+25·5	91	444	
	407	0·582	38·4	92·6	+23·5	0	8	657 f
	407	0·603	40·2	90·6	+23·8	5	30	145
		0·962	64·7			(110)	(555)	(1247)
Jan. 6								
5·992		0·982	239·2					220
	407	0·517	17·1	92·0	+25·6	61	439	
	407	0·495	22·1	90·0	+23·3	0	10	
		0·902	60·3					240
Jan. 7						(61)	(449)	(460)

Mean Solar Time.	No. of Group and Letter for Spot.	Distance from Centre in terms of Sun's Radius.	Position Angle from Sun's Axis.	HELIOGRAPHIC Longitude.	HELIOGRAPHIC Latitude.	SPOTS Area of UMBRA for each Spot (and for Day).	SPOTS Area of WHOLE for each Spot (and for Day).	FACULÆ Area for each Group (and for Day).
1881.			°	°	°			
7ᵈ·634	407	0·525	336·3	93·5	+24·6	25	91	
	407	0·520	340·7	91·0	+25·2	51	207	
Jan. 9*						(76)	(298)	
8·716	407	0·636	319·1	93·0	+24·9	30	80	
	407	0·632	320·9	91·9	+25·4	4	18	
	407	0·615	321·0	90·9	+24·6	8	27	
	407	0·622	323·7	89·9	+26·1	69	173	
Jan. 10*						(111)	(298)	
9·782	407	0·760	308·4	92·5	+24·8	31	114	
	407	0·740	311·7	89·5	+25·9	46	160	
Jan. 11*						(77)	(274)	
10·783	407	0·849	303·3	91·4	+25·0	23	58	
	407	0·830	306·0	88·4	+26·2	28	113	1108 c
Jan. 12*						(51)	(171)	(1108)
11·776	407	0·941	297·8	91·2	+24·2	9	68	
	407	0·924	300·0	87·8	+25·4	44	132	1994 c
	407a	0·856	301·0	78·4	+23·3	0	42	
Jan. 13*						(53)	(242)	(1994)
12·775	407	0·985	295·2	89·2	+23·8	17	87	
	407	0·977	297·2	86·2	+25·3	35	142	1206 n f
Jan. 14*						(52)	(229)	(1206)
13·787		0·982	298·4					400
Jan. 15*								(400)
15·763	408	0·476	320·0	351·5	+16·6	18	97	
	408	0·465	328·8	347·7	+18·6	30	113	
	409	0·403	229·4	351·9	−19·9	19	68	
	409	0·353	226·5	348·6	−18·9	0	41	
	409a	0·573	19·3	320·7	+27·7	6	26	
Jan. 17*						(73)	(345)	
16·784	408	0·632	303·5	352·7	+16·1	0	109	
	408	0·563	311·7	347·4	+18·7	25	163	
	409	0·585	241·8	352·8	−20·3	0	72	
	408	0·528	239·8	348·5	−19·9	0	100	
	409a	0·548	356·8	321·6	+28·0	0	37	
Jan. 18*						(25)	(481)	
17·886		0·951	295·8					689
	408	0·811	295·5	354·9	+16·9	39	184	
	408	0·775	296·9	351·2	+16·9	2	61	
	408	0·740	399·4	347·5	+17·5	2	46	
	408	0·750	302·3	346·5	+19·4	46	240	
	409	0·759	147·7	353·3	−20·2	14	47	
	409	0·703	247·5	348·6	−19·8	9	104	
	410a	0·558	335·9	319·7	+25·5	0	20	

* Indian photo.　　　　The Areas of Spots and Faculæ are expressed in Millionths of the Sun's visible Hemisphere.

MEASURES of POSITIONS and AREAS of SPOTS and FACULÆ upon the SUN'S DISK on PHOTOGRAPHS taken in the YEAR 1881—*continued*.

Mean Solar Time	No. of Group, and Letter for Spot	Distance from Centre in terms of Sun's Radius	Position Angle from Sun's Axis	HELIOGRAPHIC Longitude	HELIOGRAPHIC Latitude	SPOTS. Area of UMBRA for each Spot (and for Day)	SPOTS. Area of WHOLE Spot (and for Day)	FACULÆ. Area for each Group (and for Day)
1881. 17ᵈ·886	4107	0·558	33·9	317·4	+26·4	0	17	
	4100	0·563	34·5	315·5	+27·5	0	17	
	410	0·308	210·8	314·7	−20·4	0	15	
	410	0·286	200·2	311·1	−20·7	2	15	
	410b	0·957	72·5	234·4	+15·0	5	121	1153 c
		0·921	61·2					589
Jan. 19*						(119)	(887)	(2431)
18·821	408	0·919	292·0	356·0	+17·8	23	167	
	408	0·902	291·7	353·9	+17·3	0	50	
	408	0·876	293·4	350·0	+17·6	13	70	
	408	0·850	295·1	346·6	+18·1	17	87	
	408	0·852	296·7	346·0	+20·1	59	264	
	409	0·862	251·0	352·0	−19·0	0	73	
	409	0·827	250·3	348·0	−19·1	14	38	
	410a	0·622	321·0	317·5	+24·5	0	10	
	410a	0·613	326·4	314·7	+25·8	4	18	
	410	0·441	234·3	315·0	−19·6	7	41	
	410	0·422	230·6	313·0	−20·4	5	17	
	410	0·417	222·6	310·4	−22·7	0	15	
	410	0·404	230·7	312·0	−19·7	0	8	
	410	0·391	227·3	310·5	−20·2	0	8	
	410c	0·191	177·7	292·2	−16·2	0	8	
	410c	0·203	169·8	290·6	−16·7	0	11	
	410b	0·850	70·6	237·3	+13·5	0	26	
	410b	0·905	70·1	230·9	+15·5	0	26	
Jan. 20*						(142)	(933)	(990)
19·970		0·818	293·0					151
	408	0·985	289·3	355·0	+18·3	126	364	
	408	0·950	291·8	346·0	+18·8	9	77	300 c
	408	0·951	294·2	345·8	+21·0	55	204	
	409	0·940	252·4	348·1	−18·4	5	25	232 c
	410	0·651	246·1	316·6	−19·4	10	40	
	410b	0·594	242·1	311·6	−20·6	7	20	
	410b	0·697	64·6	237·5	+13·3	0	21	
	410b	0·734	64·3	234·6	+14·6	0	17	
Jan. 21						(212)	(768)	(969)
20·720	408	0·986	289·7	345·6	+18·4	0	45	
	408	0·986	292·3	345·2	+20·9	27	136	994 c
	410	0·771	248·6	317·3	−19·8	2	27	
	410	0·734	246·8	313·7	−20·6	0	11	
	410	0·702	245·7	310·8	−20·7	6	19	
	410b	0·580	57·3	237·5	+13·5	0	26	
	410b	0·621	58·0	235·0	+14·6	2	18	
	410b	0·645	60·2	231·7	+14·2	0	22	
Jan. 22*						(37)	(304)	(994)
21·659		0·917	303·6					767
	410	0·883	250·8	317·4	−19·5	0	16	850 c
	410b	0·470	44·4	235·6	+14·4	0	42	
	410a	0·442	12·6	232·8	−19·0	0	17	
	410b	0·461	121·9	231·0	−19·1	7	17	
		0·849	111·9					552
		0·987	114·4					655
Jan. 23*						(7)	(92)	(2824)

Mean Solar Time	No. of Group, and Letter for Spot	Distance from Centre in terms of Sun's Radius	Position Angle from Sun's Axis	HELIOGRAPHIC Longitude	HELIOGRAPHIC Latitude	SPOTS. Area of UMBRA for each Spot (and for Day)	SPOTS. Area of WHOLE Spot (and for Day)	FACULÆ. Area for each Group (and for Day)
1881. 22ᵈ·786		0·951	251·1					1240
	410b	0·348	12·7	236·1	+14·1	0	33	
	410d	0·232	155·6	235·1	−17·8	34	125	
	410d	0·312	146·9	230·2	−20·6	23	112	
	410e	0·603	113·9	176·3	−23·9	0	32	693 c
	410e	0·930	115·5	172·2	−25·7	0	59	
Jan. 24*						(39)	(361)	(1933)
23·675		0·888	253·4					336
	410b	0·356	336·6	237·1	+13·4	8	28	
	410d	0·246	214·6	237·2	−17·3	48	242	
	410d	0·261	205·4	235·5	−19·2	0	6	
	410d	0·219	199·8	233·2	−17·5	0	29	
	410d	0·228	189·7	231·1	−18·7	18	48	
	410d	0·249	185·5	230·3	−20·0	0	31	
	410d	0·266	182·2	229·4	−21·1	32	97	
	410e	0·813	114·8	175·5	−23·3	0	27	
	410e	0·831	115·6	173·7	−24·3	0	25	561 s.
Jan. 25*						(108)	(543)	(897)
24·807	410d	0·447	242·8	238·4	−17·0	86	405	
	410d	0·365	229·7	230·9	−19·0	17	66	
	410d	0·343	231·0	230·1	−17·9	0	7	
	410d	0·350	227·2	229·6	−19·2	18	33	
	410d	0·347	224·8	228·9	−19·7	16	31	
	410d	0·357	221·8	228·5	−20·9	99	165	
	410d	0·373	217·3	228·0	−22·7	3	17	
	410e	0·416	315·1	228·1	+14·4	0	14	
	410e	0·671	110·3	174·6	−24·2	0	15	
		0·948	105·4					288
Jan. 26*						(239)	(738)	(288)
25·672	410d	0·612	249·6	239·2	−17·0	113	464	
	410d	0·594	253·2	238·3	−14·6	0	6	
	410d	0·594	251·3	238·2	−15·1	0	6	
	410d	0·579	245·4	236·1	−18·8	2	6	
	410d	0·561	248·2	235·3	−16·9	4	9	
	410d	0·515	238·6	230·4	−20·7	9	22	
	410d	0·507	240·9	230·4	−19·3	7	18	
	410d	0·489	242·3	229·5	−18·3	0	11	
	410f	0·491	235·3	228·1	−21·4	50	129	
	410f	0·597	304·7	231·7	+14·1	11	19	
	410f	0·565	306·8	230·3	+14·6	6	15	
	410f	0·536	309·0	227·9	+14·4	24	64	
	410e	0·588	127·1	171·3	−25·7	0	4	
	410e	0·602	127·6	170·5	−26·4	8	10	
Jan. 27*						(253)	(818)	
26·675	410d	0·786	252·3	240·8	−17·5	112	473	
	410d	0·749	253·3	237·3	−16·5	0	37	
	410d	0·734	253·9	236·2	−15·8	1	14	617 c
	410d	0·716	251·5	234·5	−17·3	9	22	
	410d	0·676	244·7	230·1	−21·3	6	+29	
	410d	0·665	246·0	229·5	−20·2	2	6	

* Indian photo.   The Areas of Spots and Faculæ are expressed in Millionths of the Sun's visible Hemisphere.

MEASURES of POSITIONS and AREAS of SPOTS and FACULÆ upon the Sun's Disk on PHOTOGRAPHS taken in the

Left portion

Mean Solar Time.	No. of Group and Letter for Spot.	Distance from Centre in terms of Sun's Radius.	Position Angle from Sun's Axis.	Heliographic Longitude.	Heliographic Latitude.	Area of UMBRA for each Spot (and for Day).	Area of WHOLE Spot (and for Day).	Area for each Group (and for Day).
1881.			°	°	°			
26ᵈ·675	410d	0·658	245·0	228·8	−20·8	8	12	
	410d	0·646	242·9	227·4	−21·8	32	101	287 c
	410f	0·730	295·3	231·9	+13·8	20	32	
	410f	0·678	298·5	227·1	+14·2	27	57	
	411	0·479	41·8	170·0	+15·3	23	40	
	411	0·493	42·9	169·0	+15·6	0	25	
	411	0·510	47·0	166·7	+14·8	27	68	
	413	0·967	105·6	113·2	−16·6	6	42	238 c
		0·933	118·9					763
Jan. 28*						(274)	(958)	(1905)
27·690	410d	0·871	293·2					2499
	410d	0·912	253·0	242·3	−18·0	168	560	
	410d	0·844	255·3	233·6	−17·3	17	115	1558 c
	410d	0·790	247·1	227·3	−21·6	20	50	
	411	0·377	13·5	170·8	+15·4	72	267	
	411	0·366	20·5	168·4	+14·1	8	30	
	411	0·397	23·5	166·6	+15·4	42	134	
	412	0·190	129·0	167·3	−12·8	17	55	
	412	0·240	129·3	165·0	−14·6	8	14	
	412	0·219	120·8	164·9	−12·3	2	8	
	412	0·245	120·0	163·5	−12·8	29	64	
	412	0·269	117·8	161·9	−13·0	8	14	
	413	0·878	105·1	114·3	−16·1	32	112	573 c
	414	0·920	105·3	108·6	−16·4	0	16	
Jan. 29*						(423)	(1443)	(4630)
28·693		0·920	289·5					1581
	410d	0·975	252·3	241·1	−18·6	122	476	
	410d	0·930	253·1	231·7	−17·9	42	116	1718 c
	410d	0·921	252·2	230·4	−18·7	0	47	
	411	0·385	340·1	170·6	+15·3	80	334	
	411	0·361	348·3	167·1	+14·6	58	215	
	412	0·153	220·2	168·6	−12·7	44	122	
	412	0·150	207·8	166·9	−13·6	44	133	
	412	0·118	201·6	165·3	−13·3	20	44	
	412	0·147	195·9	165·2	−14·1	16	28	
	412	0·116	181·1	162·9	−12·5	47	129	
	412	0·145	174·9	161·0	−14·3	34	59	
	413	0·737	106·9	115·6	−16·5	23	68	
	413	0·797	109·9	110·4	−19·4	3	13	
	414	0·811	104·8	108·6	−15·4	3	13	1091 c
		0·777	129·0					684
		0·919	63·6					388
Jan. 30*						(536)	(1797)	(5462)
29·932		0·990	287·3					355
		0·984	248·2					344
		0·932	251·0					244
		0·909	234·7					270
		0·801	248·0					216
	411	0·532	311·9	170·3	+15·2	24	185	
	411	0·531	315·2	169·1	+14·6	0	1	
	411	0·502	312·0	168·7	+13·9	2	26	
	411	0·506	314·2	168·1	+14·9	0	32	
	411	0·494	317·1	166·5	+15·4	19	116	
	412	0·407	253·2	169·6	−12·4	67	260	

Right portion

Mean Solar Time.	No. of Group and Letter for Spot.	Distance from Centre in terms of Sun's Radius.	Position Angle from Sun's Axis.	Heliographic Longitude.	Heliographic Latitude.
1881.				°	°
29ᵈ·952	412	0·375	249·7	167·3	−13·2
	412	0·346	253·8	165·9	−11·3
	412	0·350	247·2	165·5	−13·8
	412	0·307	246·0	162·8	−13·0
	412	0·287	241·6	161·2	−13·7
	413	0·519	112·2	116·2	−16·6
	414	0·636	105·5	107·1	−14·5
	415	0·981	69·3	70·5	+18·8
		0·907	56·0		
Jan. 31					
30·787	411	0·934	251·2		
	411	0·664	301·9	170·9	+15·5
	411	0·643	301·0	169·7	+14·2
	411	0·630	300·3	169·1	+13·4
	411	0·628	301·6	168·5	+14·0
	411	0·611	303·3	166·7	+14·6
	411	0·613	306·0	166·1	+15·8
	412	0·584	257·5	170·7	−12·3
	412	0·563	256·8	169·2	−12·5
	412	0·557	255·1	168·7	−13·4
	412	0·525	252·3	166·4	−14·5
	412	0·494	253·0	164·2	−13·7
	412	0·486	254·1	163·8	−13·1
	412	0·469	255·2	162·7	−12·4
	412	0·468	253·1	162·5	−13·4
	412	0·455	250·5	161·4	−14·3
	413	0·354	121·9	117·0	−16·6
	415	0·924	66·0	72·0	+19·3
Feb. 1*					
32·788	412	0·889	290·9		
	412	0·888	259·8	172·0	−12·0
	412	0·807	257·1	162·9	−14·1
	412	0·793	258·1	161·6	−13·3
	412	0·782	256·4	160·5	−14·6
	413	0·230	219·9	117·7	−16·4
	415a	0·436	186·5	112·2	−31·9
	415a	0·446	182·0	110·0	−32·6
	415	0·692	54·4	72·6	+18·7
	415	0·728	57·3	68·8	+18·4
	415	0·765	57·5	65·8	+19·7
	415	0·791	57·9	63·4	+20·5
		0·754	42·6		
Feb. 3*					
33·777	412	0·965	289·3		
	412	0·969	259·8	172·1	−11·4
	412	0·908	258·6	161·5	−13·0
	412	0·901	257·4	160·5	−14·1
	413	0·409	244·2	118·2	−16·2
	415	0·555	40·7	73·3	+18·9
	415	0·615	47·9	67·0	+18·7
	415	0·651	48·5	64·5	+20·1
		0·883	63·4		
Feb. 4*					

* Indian photo.   The Areas of Spots and Faculæ are expressed in Millionths of the Sun's visible Hemisphere.

MEASURES of POSITIONS and AREAS of SPOTS and FACULÆ upon the SUN's DISK on PHOTOGRAPHS taken in the YEAR 1881—*continued.*

Mean Solar Time.	No. of Group and Letter for Spot.	Distance from Centre in terms of Sun's Radius.	Position Angle from Sun's Axis.	HELIOGRAPHIC Longitude.	Latitude.	SPOTS Area of UMBRA for each Spot (and for Day).	Area of WHOLE for each Spot (and for Day).	FACULÆ Area for each Group (and for Day).
1881.								
34d·991	416	0·675	312·6	112·9	+21·7	5	13	
	415	0·451	14·6	73·8	+19·4	83	359	
	415	0·440	19·2	72·0	+18·1	5	47	
	415	0·391	23·2	71·5	+14·6	0	11	
	415	0·461	26·1	68·4	+18·1	0	5	
	415	0·495	28·9	66·0	+19·3	0	53	
	415	0·536	29·3	64·3	+21·6	0	8	
	415	0·530	32·1	63·3	+20·4	41	135	
Feb. 5						(134)	(633)	
35·668	416	0·777	304·4	114·3	+21·4	7	12	
	415	0·439	353·3	74·0	+19·2	57	321	
	415	0·415	356·0	72·6	+17·9	9	26	
	415	0·424	358·0	71·8	+18·4	0	3	
	415	0·435	358·5	71·6	+19·2	0	3	
	415	0·418	0·8	70·5	+18·1	0	2	
	415	0·408	2·1	70·0	+17·5	10	19	
	415	0·447	13·1	64·8	+19·3	3	19	
	415	0·480	12·9	64·3	+21·4	0	4	
	415	0·467	16·3	62·9	+20·1	33	128	
	417	0·610	47·8	42·6	+18·4	0	8	
	417	0·610	49·2	42·0	+17·7	8	10	578f
	419	0·980	72·0	355·0	+16·2	77	346	1498c
		0·902	109·4					179
		0·908	120·4					220
Feb. 6*						(204)	(901)	(2475)
36·773	416	0·912	297·5	116·7	+21·7	75	187	
	416	0·888	299·9	112·8	+22·7	10	33	1385c
	416	0·876	297·9	111·9	+20·5	45	64	
	416	0·870	299·0	110·9	+21·2	9	34	
	415	0·515	325·9	74·1	+18·9	72	338	
	415	0·479	327·0	72·2	+17·2	7	18	
	415	0·491	328·7	71·9	+18·4	4	18	
	415	0·467	333·1	69·2	+18·1	9	23	
	415	0·491	341·5	66·0	+21·2	2	5	
	415	0·459	342·8	64·6	+19·4	9	26	
	415	0·457	345·3	63·4	+19·6	14	19	
	415	0·461	347·7	62·4	+20·1	17	78	
	417	0·467	26·4	43·8	+18·2	31	92	
	417	0·487	33·9	39·9	+17·5	0	5	
	417	0·481	36·0	39·3	+16·5	0	4	
	417	0·485	35·7	39·3	+16·8	7	14	
	419	0·904	69·1	355·3	+15·7	100	269	1940f
Feb. 7*						(411)	(1227)	(3325)
37·828								289
								2144c
	416	0·918	254·8	117·6	+21·5	117	335	
	416	0·981	293·6	111·2	+23·2	11	54	
	415	0·959	296·7					
	415	0·677	311·8	75·3	+21·2	0	21	
	415	0·646	310·1	74·1	+18·9	61	434	
	415	0·589	315·1	68·6	+18·7	6	13	
	415	0·553	322·8	63·4	+19·9	0	15	
	415	0·546	325·0	62·1	+20·3	15	24	
1881.								
37d·828	417	0·423	356·0	44·5	+18·3	34	68	
	417	0·414	359·8	42·8	+17·8	0	3	
	419	0·796	64·2	354·7	+15·9	76	343	
	419	0·809	66·5	351·8	+14·6	0	39	1147f
Feb. 8*						(320)	(1339)	(3580)
38·946		0·966	236·3					200
		0·886	305·7					300
	415	0·786	300·3	73·4	+18·7	90	408	354c
	415	0·684	309·2	62·0	+20·1	0	7	452u;
	417	0·490	327·3	43·8	+17·9	10	54	
	417	0·432	334·8	38·8	+16·4	0	23	
	418	0·469	349·5	33·0	+20·7	6	17	
	419	0·647	55·8	354·1	+15·8	59	271	559f
	420	0·809	62·0	339·3	+17·9	0	12	140c
	421	0·924	108·8	319·6	-19·8	0	9	497c
		0·924						341
Feb. 9						(165)	(801)	(2843)
40·046	415	0·902	294·2	73·2	+18·4	41	371	
	417	0·631	309·6	43·9	+17·9	11	36	
	418	0·583	321·4	36·1	+18·8	10	19	
	418	0·536	328·5	30·6	+20·7	0	5	
	419	0·499	40·9	353·5	+15·7	58	269	
Feb. 10						(120)	(700)	
40·625	415	0·947	292·3	73·0	+18·6	71	238	
	417	0·715	303·8	44·2	+18·2	29	95	
	418	0·660	317·6	36·8	+20·8	21	61	
	419	0·430	27·0	353·9	+15·9	82	265	
Feb. 11*						(203)	(659)	
41·975								557
	417	0·882	293·0					
	417	0·874	295·6	43·9	+18·4	21	71	
	417	0·828	295·9	38·9	+16·9	5	25	416c
	418	0·830	301·1	37·3	+21·0	10	18	224f
	419	0·395	346·4	353·4	+15·7	50	249	
	422	0·470	120·0	322·4	-19·7	11	27	
	422	0·514	123·3	320·3	-22·4	2	8	
		0·806	48·5					58
		0·969	113·7					324
Feb. 12						(99)	(398)	(1579)
42·646	417	0·929	293·4	43·1	+18·8	16	62	
	419	0·453	327·8	353·6	+15·9	66	268	
	422	0·341	130·5	323·2	-19·3	13	20	
Feb. 13*						(95)	(350)	
43·841		0·963	294·1					2045
		0·790	331·0					1063
	419	0·615	307·2	354·4	+15·9	59	247	
	422	0·227	172·8	322·2	-19·9	5	16	
	422	0·242	167·0	320·6	-20·5	0	13	
	423	0·981	68·2	248·7	+19·7	24	118	1946c
	424	0·985	109·3	242·3	-20·1	70	220	
Feb. 14*						(158)	(614)	(5054)

* Indian photo.   The Areas of Spots and Faculæ are expressed in Millionths of the Sun's visible Hemisphere.

MEASURES of POSITIONS and AREAS of SPOTS and FACULÆ upon the SUN'S DISK on PHOTOGRAPHS taken in the YEAR 1881—*continued.*

Mean Solar Time.	No. of Group, and Letter for Spot.	Distance from Centre in terms of Sun's Radius.	Position Angle from Sun's Axis.	Longitude.	Latitude.	Area of UMBRA for each Spot (and for Day).	Area of WHOLE for each Spot (and for Day).	Area for each Group (and for Day).
1881.			°	°	°			
44ᵈ·813	419	0·750	298·9	353·5	+16·2	70	211	
	423	0·932	65·3	246·4	+20·0	42	127	1781 f
	424	0·930	108·6	241·2	−19·8	92	294	
Feb. 15*						(204)	(632)	(1781)
45·861	419	0·872	293·1	353·3	+16·2	63	174	1931 p
	424a	0·682	248·5	339·0	−19·7	0	19	
	424a	0·643	245·9	335·5	−20·8	22	51	
	423	0·836	60·5	246·3	+20·0	28	127	1851
	424	0·825	109·3	241·2	−19·8	49	234	1387 f
Feb. 16*						(162)	(605)	(5169)
46·756	419	0·950	289·5	353·5	+16·0	25	123	1751 n
	424a	0·780	249·4	336·0	−20·5	47	92	1037 p
	424a	0·755	247·8	333·4	−21·2	12	24	
	423	0·735	54·0	245·9	+20·1	28	122	2332 s f
	424	0·702	111·4	241·1	−19·9	50	273	1844 f
Feb. 17*						(162)	(634)	(7264)
47·753		0·933	293·3					2054
	424a	0·917	250·8	338·7	−20·4	12	35	
	424a	0·897	250·2	335·7	−20·8	49	98	1643 f
	426b	0·438	334·1	282·8	+16·2	14	26	
	423	0·605	43·1	245·4	+19·9	33	143	991 sf
	424	0·541	117·3	240·7	−20·4	74	278	
Feb. 18*		0·954	106·0			(182)	(580)	1638 (6326)
48·761	424a	0·977	250·5	337·8	−20·5	0	29	
	424a	0·960	250·2	333·5	−20·9	16	38	912 f
	423	0·506	26·2	244·8	+20·0	27	144	
	424	0·374	128·6	240·4	−20·3	67	286	
	425a	0·852	103·2	199·6	−16·6	6	15	
	425b	0·877	103·5	196·6	−17·0	0	6	
Feb. 19*		0·861	104·8			(116)	(508)	1448 (2360)
49·770	423	0·460	1·2	244·6	+20·2	30	105	
	424	0·243	160·2	240·2	−20·3	56	258	
	425a	0·705	106·2	200·5	−16·5	11	22	
	425	0·739	104·8	197·5	−15·7	0	7	
	425b	0·753	106·2	196·3	−16·9	0	7	
	426	0·935	102·4	175·1	−14·1	31	153	1918 f
Feb. 20*						(128)	(552)	(1918)
50·909	423	0·516	333·1	244·8	+20·4	29	118	
	424	0·284	214·7	240·3	−20·5	60	261	
	426b	0·452	16·9	222·5	+18·5	0	9	
	426b	0·476	17·9	221·5	+19·8	0	7	
	426b	0·444	21·9	220·5	+17·2	14	24	
	426b	0·455	21·3	220·4	+18·0	0	4	

Mean Solar Time.	No. of Group, and Letter for Spot.	Distance from Centre in terms of Sun's Radius.	Position Angle from Sun's Axis.	Longitude.	Latitude.	Area of UMBRA for each Spot (and for Day).	Area of WHOLE for each Spot (and for Day).	Area for each Group (and for Day).
1881.			°	°	°			
50ᵈ·909	426b	0·466	21·0	220·3	+18·7	0	4	
	425m	0·481	111·0	202·6	−16·2	7	16	
	425w	0·499	111·0	201·5	−16·5	11	18	
	425m	0·502	107·8	200·9	−15·0	5	18	
	425e	0·516	110·2	200·2	−16·4	18	57	
	425b	0·569	110·1	196·5	−17·2	0	9	
	425b	0·569	107·9	196·3	−16·0	28	85	
	426	0·818	102·7	175·1	−14·4	32	94	1443 f
Feb. 21*						(194)	(724)	(1443)
51·597	423	0·581	321·3	243·9	+20·3	19	93	
	424	0·380	231·5	239·6	−20·4	41	218	
	426b	0·426	356·3	222·9	+18·0	0	16	
	426b	0·413	2·7	220·0	+17·2	0	19	
	426b	0·431	3·8	219·5	+18·3	0	14	
	425a	0·340	117·9	203·1	−15·9	0	44	
	425a	0·374	113·7	200·5	−15·3	0	5	
	425a	0·396	115·6	199·5	−16·4	0	41	
	425b	0·460	111·4	194·8	−16·0	0	55	
	426	0·729	103·4	174·3	−14·6	0	55	
Feb. 22†						(60)	(560)	
52·667	423	0·710	307·6	243·7	+19·8	33	110	
	424	0·363	242·3	239·5	−20·8	75	283	
	426b	0·499	316·6	223·6	+17·7	2	7	
	426b	0·460	330·7	220·6	+16·6	0	7	
	426b	0·449	333·7	219·0	+16·6	0	3	
	425a	0·151	165·0	204·8	−15·6	13	24	
	425a	0·179	141·5	200·5	−15·1	0	3	
	425a	0·199	144·9	200·3	−16·5	2	5	
	425a	0·235	133·7	196·9	−16·4	0	3	
	425b	0·251	128·7	195·2	−16·2	21	56	
	426	0·543	105·9	174·6	−14·6	28	80	
	426	0·548	109·6	174·6	−16·7	0	6	
Feb. 23*						(174)	(587)	
53·816		0·864	234·3					769
	423	0·885	198·6	248·1	+21·1	0	13	
	423	0·848	299·4	243·7	+20·2	29	117	1078 c
	424	0·740	248·9	239·2	−20·4	39	301	
	425a	0·295	239·0	207·1	−15·7	11	23	
	425a	0·234	230·8	202·7	−15·6	2	6	
	425b	0·166	201·0	195·4	−16·1	16	24	
	426	0·333	116·7	174·6	−15·2	25	66	
	427	0·979	65·8	117·8	+21·9	67	150	371 c
Feb. 24*						(189)	(700)	(2418)
	423	0·941	294·3	243·0	+20·0	46	155	2263 f
	424	0·877	260·5	239·1	−20·6	61	226	216c c
	425a	0·521	215·3	208·2	−15·5	15	33	
	425a	0·504	252·1	207·0	−15·1	2	7	
	425a	0·438	249·6	201·4	−15·3	7	23	
	426	0·146	165·8	175·2	−15·3	19	68	
	427	0·913	61·9	117·2	+22·0	15	53	2600
Feb. 25*						(165)	(565)	(7023)

* Indian photo.   The Areas of Spots and Faculae are expressed in Millionths of the Sun's visible Hemisphere.   † Melbourne photo.

MEASURES of POSITIONS and AREAS of SPOTS and FACULÆ upon the SUN'S DISK on PHOTOGRAPHS taken in the

Mean Solar Time.	No. of Group, and Letter for Spot.	Distance from Centre in terms of Sun's Radius.	Position Angle from Sun's Axis.	HELIOGRAPHIC Longitude.	HELIOGRAPHIC Latitude.	SPOTS Area of UMBRA for each Spot (and for Day).	SPOTS Area of WHOLE Spot (and for Day).	FACULÆ Area for each Group (and for Day).
1881.			°	°	°			
56d.012	423	0.994	291.4	243.4	+20.2	7	50	397 f
	424	0.965	250.9	239.2	−20.3	59	214	294 s f
	425a	0.716	255.4	208.8	−15.4	10	22	
	425a	0.627	254.6	201.7	−15.3	0	6	
	426	0.243	233.3	174.6	−15.4	9	19	
	427	0.811	56.2	116.6	+21.9	18	62	764 s f
		0.958	68.2					240
Feb. 26						(103)	(373)	(1695)
56.607	424	0.987	250.6	237.7	−20.3	47	217	1501 f
	427a	0.510	291.0	216.9	+15.7	11	37	1983 p
	425	0.803	256.4	208.9	−15.2	10	26	626 c
	427b	0.631	308.7	186.0	+17.0	4	8	
	427b	0.611	310.0	184.3	+16.8	0	28	
	426	0.356	245.0	174.6	−15.4	10	13	
	427	0.736	51.3	117.1	+21.7	16	32	1806 c
Feb. 27*						(98)	(361)	(5916)
57.712		0.889	256.2					507
		0.930	64.7					1050
	427	0.601	38.0	117.3	+21.6	0	14	
	427	0.642	38.0	115.2	+23.8	0	20	
Feb. 28*						(0)	(34)	(1557)
58.783		0.969	254.3					1426
		0.882	294.9					197
	427	0.502	17.2	117.4	+21.4	5	14	
	427	0.562	20.4	114.1	+24.7	8	19	869
		0.924	62.0					(2492)
Mar. 1*						(13)	(33)	
60.771		0.929	256.0					1747
	428	0.923	64.2	38.0	+20.4	20	52	2200 c
Mar. 3*						(20)	(52)	(3947)
61.684	428	0.835	60.0	38.1	+20.1	22	54	2065 c
Mar. 4*						(22)	(54)	(2065)
62.778	428	0.709	51.7	37.7	+20.2	33	88	1102 c
Mar. 5*						(33)	(88)	(1102)
63.690	428	0.596	40.0	37.9	+20.5	20	78	
Mar. 6*						(20)	(78)	
63.010		0.952	296.7		.			482
		0.928	243.5					189
		0.776	305.7					336
	428	0.480	14.3	37.2	+20.4	8	30	
	429	0.958	106.5	329.9	−17.9	0	6	602 s p
		0.881	61.2					421
Mar. 7						(8)	(36)	(2030)

Mean Solar Time.	No. of Group, and Letter for Spot.	Distance from Centre in terms of Sun's Radius.	Position Angle from Sun's Axis.	HELIOGRAPHIC Longitude.	HELIOGRAPHIC Latitude.
1881.				°	°	°
65d.918		0.983	295.5		
		0.831	306.2		
		0.802	235.8		
	428	0.473	351.5	36.7	+20.6
	429	0.869	106.6	331.7	−18.0
	430	0.988	74.7	313.8	+13.8
	431	0.989	110.8	309.1	−21.3
		0.885	59.2		
Mar. 8					
66.784		0.903	299.5		
	432	0.432	240.8	44.5	−18.7
	428	0.524	331.7	36.4	+20.4
	433	0.762	54.8	339.6	+20.6
	433	0.811	54.0	335.6	+23.5
	430	0.936	72.7	314.6	+13.4
	430	0.971	71.5	308.1	+15.9
	431	0.933	110.4	311.2	−21.6
	431	0.934	109.4	311.1	−20.6
	431	0.965	110.3	304.9	−21.4
Mar. 9*					
67.784		0.965			
		0.865	302.4		
	432	0.623	249.0	45.6	−18.6
	428	0.635	315.3	36.2	+20.4
	433	0.636	44.5	340.2	+20.1
	433	0.667	44.0	338.2	+21.1
	433	0.722	44.8	333.9	+24.9
	430	0.838	68.8	314.7	+13.4
	430	0.886	68.8	309.3	+15.0
	430	0.913	68.9	305.8	+15.9
	430	0.917	70.3	304.8	+14.8
	431	0.815	111.7	313.5	−21.8
	431	0.829	111.0	311.9	−21.4
	431	0.844	111.3	310.2	−21.8
	431	0.860	109.2	308.2	−20.2
	431	0.885	110.7	305.1	−21.7
Mar. 10*					
68.996		0.958	279.2		
		0.800	303.0		
	432	0.820	252.5	47.2	+18.5
	433	0.486	21.8	340.9	−19.7
	433	0.556	26.7	336.3	+22.8
	433	0.617	29.6	332.3	+25.6
	430	0.673	60.4	315.1	+13.7
	430	0.724	61.8	310.9	+14.6
	430	0.783	63.4	305.5	+15.6
	431	0.626	114.3	314.4	−20.7
	431	0.696	113.2	308.8	−21.2
	431	0.758	112.8	304.3	−21.9
	434	0.779	69.8	277.1	+18.0
Mar. 11					

* Indian photo.   The Areas of Spots and Faculæ are expressed in Millionths of the Sun's visible Hemisphere.

MEASURES of POSITIONS and AREAS of SPOTS and FACULÆ upon the SUN's DISK on PHOTOGRAPHS taken in the YEAR 1881—*continued.*

Mean Solar Time	No. of Group and Letter for Spot	Distance from Centre in terms of Sun's Radius	Position Angle from Sun's Axis	Longitude	Latitude	Area of UMBRA for each Spot (and for Day)	Area of WHOLE Spot (and for Day)	Area for each Group (and for Day)
1881. 69d.895								
	433	0.830	301.2					1502
	433	0.453	355.5	342.3	+19.5	62	283	
	433	0.458	3.7	338.4	+19.9	0	2	
	433	0.483	2.4	339.0	+21.5	0	2	
	433	0.483	5.1	337.6	+21.4	9	23	
	433	0.517	7.4	336.0	+23.6	15	35	
	433	0.519	8.8	335.2	+23.6	4	7	
	433	0.510	9.3	335.1	+22.9	2	5	
	433	0.554	9.5	334.4	+25.8	8	21	
	433	0.525	10.8	334.1	+23.8	6	9	
	433	0.561	12.3	332.6	+26.0	25	86	
	435	0.250	137.6	330.1	−17.7	41	94	
	435	0.500	131.0	326.7	−18.6	17	50	
	430	0.539	49.0	316.1	+13.7	13	37	
	430	0.582	53.1	311.6	+14.1	4	23	
	430	0.644	55.8	306.8	+15.2	16	31	
	431	0.465	121.9	315.3	−20.8	21	50	
	431	0.543	117.8	309.4	−20.8	6	17	
	431	0.611	116.8	304.4	−21.8	8	24	
Mar. 12*	430	0.961	104.4	265.1	−15.8	0	28	1232 c
						(257)	(827)	(2734)
70.664								
	433	0.916	297.9					1012
	433	0.521	333.2	344.9	+20.4	0	2	
	433	0.490	336.8	341.7	+19.6	73	278	
	433	0.496	343.6	338.5	+21.2	4	9	
	433	0.535	353.4	333.8	+24.8	19	84	
	433	0.548	357.7	331.3	+25.9	19	75	
	435	0.189	186.6	331.2	−17.9	32	184	
	435	0.200	155.1	324.6	−18.0	21	145	
	430	0.430	35.3	315.2	+13.6	12	19	
	430	0.510	44.1	308.5	+14.7	24	101	
	431	0.344	134.8	314.8	−20.9	25	101	
	431	0.500	122.8	303.0	−22.2	13	36	
	436	0.900	104.5	265.0	−16.2	36	108	
	436	0.928	103.7	260.9	−15.4	0	21	876 c
	436	0.930	105.5	260.5	−17.1	17	59	
Mar. 13*	437	0.970	77.2	255.9	+10.5	32	172	1081 c
						(327)	(1394)	(2969)
71.964								
	433	0.644	313.5	342.5	+20.1	73	311	341 p
	433	0.609	332.8	330.8	+25.9	13	93	153 s p
	435	0.376	237.5	332.2	−18.3	49	234	
	435	0.289	227.1	325.7	−18.3	9	95	
	430	0.357	358.9	313.2	+13.7	11	28	
	431	0.243	188.0	314.9	−21.0	8	130	
	431	0.334	149.1	302.1	−23.6	6	35	
	434	0.682	53.7	277.7	+17.9	0	10	156 c
	436	0.714	105.2	267.3	−15.8	20	114	
	436	0.747	106.6	264.5	−17.1	24	143	
	436	0.795	106.2	260.0	−17.2	12	152	465 c
	437	0.863	78.7	255.7	+10.2	33	176	611 nf
		0.945	110.5					161
Mar. 14						(258)	(1521)	(1887)

Mean Solar Time	No. of Group and Letter for Spot	Distance from Centre in terms of Sun's Radius	Position Angle from Sun's Axis	Longitude	Latitude	Area of UMBRA for each Spot (and for Day)	Area of WHOLE Spot (and for Day)	Area for each Group (and for Day)
1881. 73d.016								
	433	0.778	304.0	342.4	+20.6	57	313	466 c
	433	0.710	319.2	330.0	+26.4	8	35	189 n p
	435	0.573	247.7	332.8	−18.5	42	226	
	435	0.456	245.4	324.6	−17.3	0	36	
	431	0.373	230.7	316.8	−20.4	28	84	
	430	0.425	330.3	311.5	+14.6	18	79	
	436	0.512	108.3	268.8	−15.4	38	135	
	436	0.562	109.8	255.5	−16.9	37	187	
	436	0.631	107.6	260.3	−16.6	36	260	1321 c
	437	0.734	68.6	255.1	+10.4	36	218	884 f
		0.859	112.3					290
Mar. 15						(300)	(1573)	(3150)
73.930								
	433	0.884	298.1	343.3	+20.8	50	220	405 c
	434	0.808	309.8	330.5	+26.0	0	2	476 p
	435	0.726	250.6	333.1	−18.9	53	211	342 c
	435	0.632	250.4	324.7	−17.6	2	28	
	430	0.584	306.4	315.8	+14.0	4	7	
	430	0.516	315.9	309.2	+15.6	0	19	
	431	0.533	242.8	317.2	−20.2	16	53	
	431	0.431	230.7	308.0	−22.4	0	6	
	431	0.392	220.6	303.1	−24.1	0	28	
	436	0.325	117.7	269.7	−15.4	59	301	
	436	0.412	110.0	264.1	−16.5	52	294	
	436	0.485	114.9	259.5	−18.0	35	166	
	437	0.579	61.1	256.1	+10.2	37	141	264 f
		0.815	63.7					818
		0.818	113.8					497
Mar. 16						(308)	(1476)	(2802)
74.695								
	433	0.716	300.5					314
	436	0.965	293.8	343.8	+20.7	37	179	838 f
	435	0.860	252.3	333.0	−18.8	32	193	564 f
	431	0.702	248.9	317.2	−19.7	9	41	
	431	0.576	242.0	306.3	−21.6	10	20	
	436	0.151	165.2	271.0	−15.4	73	302	
	436	0.220	137.1	264.3	−16.2	36	157	
	436	0.309	131.6	259.2	−18.5	33	120	
	437	0.411	45.0	256.2	+10.1	22	146	
		0.803	56.3					161
		0.811	117.5					129
Mar. 17						(252)	(1158)	(2006)
76.044								
		0.970	300.8					401
		0.820	295.5					533
	435	0.958	252.0	333.5	−19.2	49	192	476 f
	431	0.849	251.3	317.4	−19.6	4	20	
	431	0.774	246.8	309.1	−22.3	8	10	505 f
	438	0.503	303.3	284.2	+ 9.6	5	14	
	437	0.296	7.9	256.7	+10.0	14	89	
	436	0.261	234.1	271.7	−15.6	65	388	
	435	0.187	208.5	264.4	−16.1	32	136	
	436	0.191	177.2	258.5	−18.0	15	103	
		0.867	109.3					248
		0.945	67.2					431
Mar. 18						(192)	(952)	(2594)

* Indian photo.　　The Areas of Spots and Faculæ are expressed in Millionths of the Sun's visible Hemisphere.

MEASURES of POSITIONS and AREAS of SPOTS and FACULÆ upon the SUN's DISK on PHOTOGRAPHS taken in the YEAR 1881—*continued.*

Left half

Mean Solar Time	No. of Group, and Letter for Spot	Distance from Centre in terms of Sun's Radius	Position Angle from Sun's Axis	Longitude	Latitude	Area of UMBRA for each Spot (and for Day)	Area of WHOLE Spot (and for Day)	Area for each Group (and for Day)
1881. 77ᵈ·048	431	0·925	292·3					427
	431	0·947	252·3	318·1	−19·0	0	5	
	436	0·464	248·6	272·4	−16·0	72	381	
	436	0·361	240·9	265·0	−16·7	10	89	
	436	0·284	226·3	258·2	−18·1	14	97	
	437	0·351	326·2	257·2	+10·0	6	44	
	439	0·498	44·8	224·7	+14·1	45	157	
	440	0·982	117·4	164·7	−28·2	0	45	185 f
		0·740	112·3					521
		0·859	62·5					213
Mar. 19		0·982	74·7			(147)	(818)	193 (2125)
77·723	440a	0·926	249·1					1277
	436	0·889	307·6	290·2	+28·8	10	17	
	436	0·584	252·4	272·1	−15·9	58	340	
	436	0·505	249·4	266·3	−16·3	7	55	
	436	0·403	241·4	258·6	−17·7	0	57	
	437	0·451	308·7	257·7	+ 9·8	18	67	
	439	0·396	23·3	277·7	+14·3	77	334	
Mar. 20*	439	0·420	31·9	223·7	+14·0	75 (245)	243 (1173)	(1277)
78·803	436	0·980	244·4					776
	436	0·766	255·0	272·8	−15·9	68	318	
	436	0·739	254·4	270·3	−16·2	5	14	632 c
	436	0·664	252·3	263·9	−16·9	2	6	
	437	0·630	294·1	258·2	+ 9·3	4	20	
	439	0·371	349·4	226·7	+14·4	85	424	
	439	0·357	359·9	222·7	+13·9	68	338	
	440b	0·731	102·7	175·6	−14·0	0	23	
		0·854	67·7					1489
Mar. 21*		0·906	122·1			(232)	(1143)	885 (3782)
80·143		0·951	293·7					79
		0·828	287·8					367
		0·645	301·3					143
	436	0·922	255·3	273·0	−16·2	113	343	
	436	0·893	255·7	268·9	−15·9	0	37	
	436	0·860	254·3	264·8	−17·0	9	44	1383
	4·9	0·503	315·4	226·3	+14·5	100	468	
	439	0·456	321·6	221·9	+14·3	95	451	
	441	0·653	70·3	166·9	+ 7·3	8	31	
		0·759	123·8					266
Mar. 22		0·858	108·2			(325)	(1374)	204 (2442)
80·675	436	0·886	286·6					1501
	436	0·959	255·2	272·6	−16·2	55	208	
	436	0·912	255·0	264·4	−16·5	4	19	2205 c
	439	0·580	307·3	226·2	+14·5	87	425	
	439	0·523	312·0	221·9	+14·1	83	379	
Mar. 23*	441	0·544	54·3	171·1	+12·3	19 (248)	56 (1087)	(3706)

Right half

Mean Solar Time	No. of Group, and Letter for Spot	Distance from Centre in terms of Sun's Radius	Position Angle from Sun's Axis	Longitude	Latitude	Area of UMBRA for each Spot (and for Day)	Area of WHOLE Spot (and for Day)	Area for each Group (and for Day)
1881. 81ᵈ·786		0·978	282·7					1047
		0·964	252·4					617
		0·909	246·6					1060
	439	0·749	297·1	226·8	+15·0	96	275	1775 p
	439	0·679	298·6	221·0	+13·6	76	400	
	441	0·384	257·6	174·6	+13·9	0	34	
	441	0·393	31·2	171·3	+12·9	17	57	
	441	0·459	33·7	168·0	+15·9	4	18	
Mar. 24*		0·933	62·0			(193)	(784)	654 (5153)
82·905	439	0·838	246·5					317
	439	0·872	292·1	225·3	+15·5	33	222	
	439	0·830	290·9	221·1	+13·1	75	386	674 c
	441	0·308	355·2	169·9	+12·9	5	32	
	441	0·390	7·2	165·7	+15·9	5	9	
		0·867	58·2					326
Mar. 25		0·923	104·8			(118)	(649)	113 (1430)
83·555	439	0·939	290·2	226·4	+16·3	0	139	1017 c
	439	0·903	287·4	221·7	+12·5	37	204	
Mar. 26†	441	0·386	333·4	170·1	+13·5	0 (37)	24 (367)	(1017)
84·751 Mar. 27*	439	0·979	284·5	220·3	+12·7	76 (76)	191 (191)	
85·957		0·959	293·3					122
		0·938	251·5					118
		0·755	299·2					195
	442	0·844	108·3	70·6	−19·0	8	14	302 f
Mar. 28	443	0·888	66·6	70·0	+17·3	8 (16)	55 (69)	334 n p (1071)
87·016	444	0·728	233·5					178
	444	0·868	291·1	171·0	+14·6	2	7	494 n
	442	0·695	110·9	71·1	−19·1	6	17	301 f
Mar. 29	443	0·777	60·7	69·2	+17·7	16 (24)	95 (119)	607 f (1580c)
87·910		0·939	290·8					576
		0·893	247·3					411
	442	0·548	115·5	71·1	−19·2	0	42	
	443	0·666	52·1	69·0	+18·7	6	59	
		0·785	114·0					138
Mar. 30		0·815	52·7			(6)	(101)	137 (1262)
88·964		0·978	288·3					118
		0·967	248·7					113
		0·930	240·5					170
	445	0·439	169·6	83·3	−31·9	28	61	

* Indian photo.   The Areas of Spots and Faculæ are expressed in Millionths of the Sun's visible Hemisphere.   † Melbourne photo.

MEASURES of POSITIONS and AREAS of SPOTS and FACULÆ upon the SUN's DISK on PHOTOGRAPHS taken in the YEAR 1881—*continued.*

Left half

Mean Solar Time	No. of Group, and Letter for Spot	Distance from Centre in terms of Sun's Radius	Position Angle from Sun's Axis	Heliographic Longitude	Latitude	Area of UMBRA for each Spot (and for Day)	Area of WHOLE Spot (and for Day)	Area for each Group (and for Day)
1881.								
88ᵈ·964	445	0·495	163·1	78·7	−34·5	8	76	
	442	0·349	131·3	72·6	−19·5	0	7	
	443	0·536	37·2	68·7	+19·1	0	26	
		0·919	62·3					215
Mar. 31						(36)	(170)	(606)
8d·908		0·973	240·5					187
		0·922	249·8					155
		0·750	312·5					135
	446	0·818	243·8	130·0	−25·0	0	11	
	445	0·436	195·4	84·0	−31·1	26	192	
	445	0·477	181·4	77·0	−34·8	22	147	
	442	0·236	164·7	72·4	−19·5	0	21	
	447	0·936	72·1	6·0	+15·0	0	27	400 p
	448	0·969	110·6	359·4	−21·5	23	95	256 f
Apr. 1						(71)	(493)	(1133)
91·019		0·921	245·7					209
		0·844	301·2					365
	449	0·664	304·4	96·3	+16·8	0	21	
	445	0·537	219·5	84·7	−30·2	31	189	
	445	0·531	201·5	75·3	−35·6	36	150	
	447	0·854	68·0	6·7	+15·1	2	6	371 f
	448	0·880	110·3	359·8	−20·9	18	91	595 f
Apr. 2						(87)	(457)	(1540)
91·754		0·883	227·3					775
	445	0·635	212·9	84·6	−30·8	47	164	
	445	0·589	217·9	74·8	−35·4	28	104	
	448	0·790	111·1	0·1	−20·5	10	65	848 f
	451	0·976	66·5	338·3	+21·2	21	100	697 p
Apr. 3*						(106)	(433)	(2320)
93·084		0·830	302·0					597
	445	0·796	236·9	84·3	−29·8	71	226	
	445	0·725	224·7	73·1	−35·8	24	96	617 c
	450	0·340	233·5	50·9	−17·6	0	59	
	448	0·589	115·3	0·0	−19·7	9	26	498 f
	451	0·887	61·9	337·3	+21·4	17	126	395 f
		0·749	51·0					120
Apr. 4						(121)	(533)	(2227)
93·785		0·963	288·5					435
		0·850	298·4					946
	445	0·875	239·1	84·1	−29·8	31	232	
	445	0·798	228·7	72·6	−36·0	11	39	1070 c
	450	0·472	242·8	51·1	−18·0	27	65	
	448	0·470	121·3	359·9	−19·7	9	14	
	451	0·813	57·6	337·5	+21·7	27	95	848 f
		0·929	108·6					1039
Apr. 5*						(106)	(445)	(4338)

Right half

Mean Solar Time	No. of Group, and Letter for Spot	Distance from Centre in terms of Sun's Radius	Position Angle from Sun's Axis	Heliographic Longitude	Latitude	Area of UMBRA for each Spot (and for Day)	Area of WHOLE Spot (and for Day)	Area for each Group (and for Day)
1881.								
95ᵈ·009		0·915	297·3					515
	445	0·965	241·0	84·6	−29·6	39	248	409 s f
	450	0·716	251·4	54·1	−17·6	7	51	230 p
	451	0·675	47·9	336·4	+21·8	19	96	370 f
	452	0·956	112·6	295·1	−23·3	15	51	433 p
		0·867	66·2					529
Apr. 6						(80)	(446)	(2286)
95·925		0·973	235·7					387
		0·929	299·5					358
	450	0·840	252·1	53·9	−18·3	0	16	247 c
	451	0·575	35·5	335·7	+22·2	20	62	
	452	0·871	113·9	296·6	−23·7	0	55	
	453	0·996	105·5	270·5	−15·9	0	401	
		0·747	39·3					185
		0·813	63·8					215
Apr. 7						(20)	(534)	(2099)
97·013		0·924	253·5					308
	451	0·486	13·7	335·3	+22·2	11	57	
	452	0·734	116·6	297·0	−23·3	8	42	128 s f
	454	0·963	60·2	273·5	+26·6	0	31	212 c
	453	0·945	104·9	270·9	−16·0	52	224	540 s
		0·985	77·0					425
Apr. 8						(71)	(354)	(1613)
98·012		0·984	253·5					240
		0·968	293·7					72
	451	0·483	348·7	335·1	+22·3	18	56	
	452	0·548	122·6	299·5	−22·2	9	14	
	452	0·628	120·3	293·3	−23·2	8	38	
	454	0·872	54·9	276·5	+26·6	0	19	180 f
	453	0·847	104·5	271·2	−15·3	38	290	710 e
		0·931	73·8					473
Apr. 9						(73)	(417)	(1675)
98·722	451	0·528	333·0	334·9	+22·4	11	32	
	452	0·511	126·2	293·5	−22·7	0	9	
Apr. 10*	453	0·753	105·1	271·2	−15·2	34	182	
						(45)	(223)	
99·809	451	0·645	316·4	334·5	+22·8	17	29	
	454a	0·548	118·6	274·9	−20·1	0	13	
Apr. 11*	453	0·570	108·1	271·6	−15·0	29	162	
						(46)	(204)	
100·730		0·839	245·7					882
	451	0·753	307·1	333·9	+22·7	5	17	316 e
	454a	0·369	130·1	276·1	−19·2	2	9	
	453	0·399	114·7	271·5	−14·8	37	202	
	455	0·568	124·8	262·9	−23·8	0	14	
	455	0·612	125·6	260·0	−25·6	8	14	
	456	0·966	74·7	220·5	+13·1	0	18	1152 e
		0·884	113·2					836
Apr. 12*						(52)	(274)	(3186)

* Indian photo.　　　The Areas of Spots and Faculæ are expressed in Millionths of the Sun's visible Hemisphere.

MEASURES of POSITIONS and AREAS of SPOTS and FACULÆ upon the SUN'S DISK on PHOTOGRAPHS taken in the YEAR 1881—*continued.*

Mean Solar Time.	No. of Group, and Letter for Spot.	Distance from Centre in terms of Sun's Radius.	Position Angle from Sun's Axis.	Longitude.	Latitude.	Area of UMBRA for each Spot (and for Day).	Area of WHOLE for each Spot (and for Day).	Area for each Group (and for Day).
1881. 101d.911		0.816	249.7					497
	451	0.816	249.7	333.1	+23.1	0	26	569 s f
	453	0.195	144.7	271.1	−14.7	37	182	
	455	0.377	146.5	264.7	−23.7	20	34	
	455	0.453	140.3	259.1	−25.6	2	6	374 n
	456	0.878	71.4	219.2	+13.4	19	90	(1440)
Apr. 13						(78)	(338)	
102.728		0.885	306.1					1769
		0.882	247.4					728
	453	0.170	204.8	271.1	−14.3	39	163	
	455	0.311	174.4	265.0	−23.5	17	68	
	455	0.345	163.1	260.6	−24.7	17	48	817 n
	456	0.788	67.7	218.4	+13.8	16	86	446
		0.947	69.8					(3760)
Apr. 14*						(89)	(365)	
103.672		0.963	305.1					823
		0.908	252.3					342
	457	0.599	247.9	290.0	−17.5	48	156	
	453	0.324	241.1	271.5	−14.1	27	153	
	455	0.361	212.6	266.7	−23.0	17	83	
	455	0.345	197.1	261.0	−24.5	0	41	1095
	456	0.460	60.8	219.2	+14.0	15	34	(2260)
		0.942	65.9					
Apr. 15*						(107)	(467)	
104.698	457	0.777	252.1	291.6	−17.3	38	153	
	453	0.521	251.7	271.5	−14.0	28	160	
	455	0.495	233.5	266.3	−22.0	33	163	
	455	0.447	223.1	260.5	−24.1	0	68	
	456	0.482	46.6	219.9	+14.2	9	18	1064
		0.908	63.0					(1064)
Apr. 16*						(108)	(562)	
105.699	457	0.906	252.9	292.9	−17.7	38	132	
	457a	0.743	249.8	274.8	−18.5	48	144	
	453	0.587	255.6	270.8	−13.7	55	150	
	455	0.647	241.5	265.4	−22.2	32	122	
	455	0.597	236.1	260.4	−23.8	0	100	
		0.925	105.9					693
Apr. 17*						(173)	(648)	(693)
107.013		0.906	251.0					595
		0.862	278.7					323
	457	0.994	252.4	295.2	−18.0	0	59	494 f
	454	0.949	299.9	276.7	+26.2	4	28	353 n f
	453	0.868	257.4	270.7	−13.4	38	184	377 c
	455	0.845	246.7	267.2	−22.3	26	134	208 r
	455	0.787	243.1	260.4	−24.2	14	33	167 c
	458	0.367	212.6	222.7	−23.0	0	5	
	458	0.347	206.5	220.0	−23.1	0	12	
	458	0.290	189.0	213.2	−21.7	0	4	
	459	0.617	45.9	182.2	+20.8	30	85	
1881. 107d.013	460	0.732	108.3	164.0	−16.8	12	19	
	460	0.776	106.6	159.9	−16.0	0	18	140 c
	460	0.818	107.9	155.9	−17.5	4	19	
	461	0.906	79.4	146.7	+7.3	15	136	311 f
	462	0.925	97.7	142.5	−9.0	0	8	105 s
	463	0.979	116.1	131.2	−26.6	0	30	305 s
		0.770	63.3					221
Apr. 18						(143)	(774)	(3599)
107.700	454	0.981	297.7	276.1	+25.9	0	81	1197 c
	453	0.933	257.5	276.6	−13.5	33	123	1328 s
	455	0.906	247.8	265.0	−22.2	26	124	
	455	0.860	244.7	259.6	−24.3	9	31	1489 c
	459	0.499	31.9	185.1	+20.1	13	56	
	459	0.531	36.6	181.7	+20.3	47	142	
	460	0.621	110.7	164.1	−16.8	8	14	
	460	0.655	108.6	161.3	−15.9	4	11	
	460	0.713	109.6	156.7	−17.5	9	23	
	461	0.817	77.5	148.6	+7.1	42	167	
	461	0.870	76.7	142.5	+8.0	23	91	656 c
	462	0.848	98.9	143.2	−10.2	0	3	121
	463	0.939	116.1	131.5	−26.2	0	23	1067 c
Apr. 19*						(225)	(889)	(6058)
108.785	455	0.970	247.3	263.5	−23.2	0	194	1598 c
	464	0.175	171.1	185.3	−14.9	5	31	
	459	0.428	5.8	184.3	+20.1	89	249	
	459	0.466	13.4	180.2	+21.9	35	87	
	460	0.486	116.3	159.9	−16.9	0	18	
	460	0.534	115.2	156.6	−17.5	19	57	
	461	0.633	72.6	149.6	+6.9	50	291	
	461	0.730	71.4	142.5	+9.9	47	117	759 c
	462	0.703	103.0	142.5	−12.7	0	16	
		0.849	116.7					1452
Apr. 20*						(245)	(1060)	(3809)
109.727	458	0.752	243.7	221.4	−22.8	10	29	
	459	0.447	336.8	185.3	+19.4	18	104	
	459	0.447	346.6	180.9	+20.8	18	77	
	464	0.265	228.5	186.4	−14.8	0	5	
	464	0.223	211.3	181.5	−15.8	0	8	
	460	0.308	132.5	160.9	−16.7	30	118	
	460	0.346	134.0	159.4	−18.6	7	17	
	460	0.366	127.7	157.0	−17.6	23	78	
	462c	0.441	103.2	148.9	−10.2	9	65	
	461	0.449	63.1	150.9	+7.2	41	207	
	461	0.527	69.1	145.0	+6.6	2	8	
	461	0.577	64.9	142.7	+10.0	20	59	
Apr. 21*						(178)	(546)	
110.703	458	0.883	246.9	223.0	−22.6	17	119	
	458	0.863	242.8	220.6	−21.5	13	44	
	458	0.854	245.8	219.3	−23.1	31	117	
	459	0.559	317.5	185.2	+19.9	10	29	
	459	0.529	321.7	181.9	+20.0	0	72	
	460	0.219	189.1	163.7	−17.2	21	124	

* Indian photo.   The Areas of Spots and Faculæ are expressed in Millionths of the Sun's visible Hemisphere.

MEASURES of POSITIONS and AREAS of SPOTS and FACULÆ upon the SUN'S DISK on PHOTOGRAPHS taken in the YEAR 1881—*continued.*

Mean Solar Time	No. of Group, and Letter for Spot	Distance from Centre in terms of Sun's Radius	Position Angle from Sun's Axis	Heliographic Longitude	Heliographic Latitude	Area of UMBRA for each Spot (and for Day)	Area of WHOLE Spot (and for Day)	Area for each Group (and for Day)
1881.								
110·d703	460	0·220	176·2	160·7	−17·5	3	46	
	460	0·240	164·0	157·6	−18·1	31	141	
	461	0·251	36·2	153·0	+ 6·9	46	141	
	461	0·398	50·9	143·4	+ 9·9	18	40	
Apr. 22°						(190)	(873)	
111·932								322
	458	0·971	288·8					
	458	0·964	247·2	220·3	−23·2	5	82	217 c
	459	0·722	303·6	185·0	+19·9	5	24	603 c
	460	0·383	235·9	164·7	−16·8	30	193	
	460	0·324	128·1	159·9	−17·0	17	59	
	460	0·295	119·1	156·6	−17·8	33	119	
	461	0·250	323·0	154·1	+ 6·8	28	147	
	461	0·251	334·4	151·7	+ 8·3	19	53	
	461	0·212	349·3	147·7	+ 7·5	4	10	
	461	0·263	9·5	142·9	+10·4	11	44	
		0·850	54·2					367
		0·944	122·0					170
Apr. 23						(152)	(731)	(1679)
113·754	465	0·887	294·4	179·9	+19·2	0	122	1851 n
	465	0·859	297·5	175·7	+20·9	41	214	
	460	0·695	151·0	164·4	−16·4	0	90	
	460	0·635	246·4	158·9	−18·3	0	17	
	460	0·600	245·9	156·3	−17·8	10	57	
	461	0·587	288·4	155·3	+ 6·9	46	180	
	461	0·563	289·8	153·4	+ 7·1	0	31	
	461	0·533	293·7	151·3	+ 6·9	8	54	
	461	0·509	294·3	149·1	+ 8·1	0	9	
Apr. 25°						(105)	(784)	(1851)
115·055		0·893	290·5					121
	465	0·958	293·4	174·1	+20·9	0	149	719 s
	465	0·849	253·0	161·9	−16·8	41	214	474 c
	461	0·800	281·2	155·9	+ 7·0	70	304	
	461	0·750	283·1	151·3	+ 6·8	5	41	
	461	0·708	285·0	147·5	+ 7·3	16	104	313 c
	465	0·363	19·6	111·4	+15·5	5	26	
		0·806	133·5					252
Apr. 26						(135)	(838)	(1899)
115·793		0·962	288·1					676
	460	0·949	254·7	166·2	−15·9	0	102	
	460	0·905	253·7	159·2	−16·6	8	95	1438 c }
	460	0·901	251·3	158·5	−17·8	0	45	
	461	0·897	279·8	157·1	+ 6·9	91	292	
	461	0·860	280·5	152·6	+ 6·8	10	35	1580 c
	461	0·815	282·3	147·6	+ 7·4	20	78	
Apr. 27°						(129)	(647)	(3694)
116·899		0·978	253·2					697
		0·786	239·5					347
	461	0·984	277·3	158·7	+ 6·4	46	247	
	461	0·960	277·5	152·6	+ 6·0	0	58	
	461	0·932	279·4	147·4	+ 7·1	44	312	572 n f
Apr. 28						(90)	(617)	(1616)

Mean Solar Time	No. of Group, and Letter for Spot	Distance from Centre in terms of Sun's Radius	Position Angle from Sun's Axis	Heliographic Longitude	Heliographic Latitude	Area of UMBRA for each Spot (and for Day)	Area of WHOLE Spot (and for Day)	Area for each Group (and for Day)
1881.								
117·d807	461	0·984	278·9	146·5	+ 8·0	0	160	1021 c
		0·956	110·6					384
		0·986	66·5					379
Apr. 29°						(0)	(160)	(1784)
118·920		0·950	250·5					206
		0·946	240·5					307
		0·826	305·7					306
	466	0·907	64·0	352·1	+21·4	12	30	273 f
		0·875	111·7					212
Apr. 30						(12)	(30)	(1304)
119·742	466	0·816	60·5	352·8	+21·1	6	36	
	466a	0·872	58·9	347·2	+24·5	10	43	480 c
		0·974	240·6					282
		0·851	304·3					501
May 1°						(16)	(79)	(1263)
120·762	466	0·696	52·6	352·3	+21·9	0	34	
May 2°						(0)	(34)	
121·898		0·888	302·7					224
		0·837	251·7					108
		0·807	229·2					352
	467	0·985	113·2	293·2	−23·5	0	60	
		0·849	54·0					199
May 3						(0)	(60)	(883)
122·756	467	0·939	113·4	293·0	−23·2	37	94	
		0·953	113·3					1426
May 4						(37)	(94)	(1426)
124·046		0·929	230·8					288
	468	0·686	256·6	28·1	−11·7	2	9	
	468	0·637	253·3	24·2	−12·0	0	2	
	469	0·317	196·1	350·7	−21·1	2	15	
	469	0·335	185·9	347·5	−22·9	7	22	
	470	0·467	14·3	338·1	+23·3	0	2	
	470	0·486	16·8	336·5	+24·1	0	4	
	470	0·512	21·8	333·3	+24·9	0	5	
	467	0·814	116·1	292·9	−23·0	23	97	
	471	0·966	103·2	270·1	−13·7	16	61	
		0·607	51·8					80
		0·824	41·7					67
		0·866	113·7					596
		0·963	76·8					316
May 5						(50)	(217)	(1357)
124·902		0·975	233·7					309
		0·798	259·0					133
	469	0·409	223·9	351·5	−20·3	0	14	
	469	0·391	213·1	347·3	−22·4	0	14	

° Indian photo. The Areas of Spots and Faculæ are expressed in Millionths of the Sun's visible Hemisphere.
‡ Taken as one spot on April 26, Greenwich.

MEASURES of POSITIONS and AREAS of SPOTS and FACULÆ upon the SUN's DISK on PHOTOGRAPHS taken in the YEAR 1881—*continued.*

Mean Solar Time.	No. of Group, and Letter for Spot.	Distance from Centre in terms of Sun's Radius.	Position Angle from Sun's Axis.	HELIOGRAPHIC Longitude.	HELIOGRAPHIC Latitude.	SPOTS Area of UMBRA for each Spot (and for Day).	SPOTS Area of WHOLE for each Spot (and for Day).	FACULÆ Area for each Group (and for Day).
1881.								
124^d·902	467	0·696	120·2	293·3	-23·0	19	97	286 *f*
	471	0·898	103·9	270·3	-13·9	31	71	504 *s f*
		0·921	74·0					626
		0·944	57·2					164
May 6						(50)	(196)	(2022)
125·894		0·917	258·5					137
	469	0·533	232·1	347·7	-22·0	2	9	
	469	0·519	228·3	345·7	-23·1	0	4	
	470	0·530	328·2	338·5	+23·5	4	6	
	470	0·531	334·6	335·4	+25·4	0	7	
	470	0·510	336·2	333·9	+24·5	0	12	
	467	0·549	128·5	293·2	-22·8	10	73	300 *f*
	471	0·776	105·2	270·6	-13·8	11	49	562 *s f*
		0·845	69·5					283
May 7						(27)	(160)	(1282)
126·707	469	0·685	242·6	350·5	-20·8	0	20	
	469	0·684	240·3	348·3	-21·7	15	37	
	470	0·629	315·5	338·8	+23·7	0	19	
	470	0·590	321·7	333·7	+24·5	10	51	
	467	0·437	141·4	292·9	-23·0	11	55	
	471	0·654	107·8	270·3	-14·0	11	32	
		0·839	117·3					404
May 8						(47)	(214)	(404)
127·983	469	0·836	246·3	348·3	-21·4	0	36	227 *p*
	470	0·773	305·1	337·0	+24·1	0	26	
	470	0·739	308·7	332·6	+25·1	19	100	433 *s p*
	467	0·338	179·4	293·1	-22·7	9	31	
	471	0·422	116·6	270·5	-13·7	4	17	
	471	0·505	124·8	267·3	-19·4	13	71	
		0·924	130·3					154
		0·934	69·5					201
May 9						(45)	(281)	(1065)
128·821		0·938	297·0					608
	469	0·911	247·3	346·8	-21·9	0	35	842 *p*
	470	0·829	302·8	332·0	+24·7	29	67	269 *p*
	472a	0·406	346·7	287·8	+20·2	9	18	
	471	0·282	131·2	269·5	-13·6	10	22	
	471	0·377	140·9	267·5	-19·9	4	16	
	472	0·283	76·6	204·0	+12·5	35	113	
		0·897	119·5					442
		0·913	68·1					419
May 10						(77)	(271)	(2580)
129·935		0·945	297·7					155
	469	0·982	247·7	346·5	-22·4	0	53	210 *c*
	472	0·895	73·5	205·8	+13·4	31	124	
	472	0·929	74·2	200·8	+13·5	0	44	
	472	0·949	73·3	197·6	+14·8	8	67	398 *c*
May 11						(39)	(288)	(763)

Mean Solar Time.	No. of Group, and Letter for Spot.	Distance from Centre in terms of Sun's Radius.	Position Angle from Sun's Axis.	HELIOGRAPHIC Longitude.	HELIOGRAPHIC Latitude.	SPOTS Area of UMBRA for each Spot (and for Day).	SPOTS Area of WHOLE for each Spot (and for Day).	FACULÆ Area for each Group (and for Day).
1881.								
130^d·987	472	0·764	70·2	206·1	+13·1	53	241	
	472	0·822	71·9	200·4	+13·1	0	19	
	472	0·850	70·7	197·7	+14·8	10	73	608 *f*
May 12						(63)	(333)	(608)
131·903		0·936	308·0					228
		0·763	240·0					310
	472	0·618	64·0	206·7	+13·5	41	188	
	472	0·685	67·1	201·1	+13·4	0	23	
	472	0·729	66·5	197·8	+15·0	18	49	583 *f*
	473	0·869	67·4	184·0	+18·5	5	10	1129 *f*
	474	0·970	83·4	166·1	+ 5·7	14	28	267 *f*
		0·965	108·5					568
May 13						(78)	(298)	(3085)
132·972		0·880	240·8					174
	472	0·437	51·0	206·9	+13·6	37	182	
	472	0·565	57·9	197·6	+15·2	0	31	
	474	0·875	81·5	167·0	+ 6·2	0	12	478 *f*
		0·800	64·0					372
		0·883	107·8					267
May 14						(37)	(225)	(1291)
133·663		0·921	240·8					189
	472	0·335	34·1	207·0	+13·7	31	190	
	474	0·783	79·7	167·5	+ 6·5	5	24	
		0·869	108·7					428
		0·888	79·7					1133
May 15ª						(36)	(214)	(1750)
134·766		0·915	291·3					364
		0·908	146·5					563
	472	0·282	347·6	207·1	+13·7	31	193	
	474	0·598	75·7	168·0	+ 6·6	6	20	
		0·846	76·2					428
May 16•						(37)	(213)	(1355)
135·700		0·924	249·4					1060
	472	0·387	316·0	207·2	+14·0	32	142	
		0·883	105·7					347
May 17•						(32)	(142)	(1407)
137·055		0·895	241·2					223
		0·790	300·3					413
	472	0·605	296·6	207·0	+14·0	31	148	
	475	0·461	332·7	186·4	+22·1	0	29	
		0·818	127·7					247
May 18						(31)	(177)	(893)
137·941		0·846	240·2					235
	476	0·859	300·7	215·8	+24·8	8	21	479 *s*
	472	0·742	291·5	207·0	+14·2	37	157	261 *f*
	475	0·571	315·7	187·0	+22·3	3	8	
	477	0·521	201·5	174·3	-30·7	11	35	

• Indian photo.   The Areas of Spots and Faculæ are expressed in Millionths of the Sun's visible Hemisphere.

MEASURES of POSITIONS and AREAS of SPOTS and FACULÆ upon the SUN's DISK on PHOTOGRAPHS taken in the YEAR 1881—*continued.*

Mean Solar Time	No. of Group, and Letter for Spot	Distance from Centre in terms of Sun's Radius	Position Angle from Sun's Axis	Longitude	Latitude	Area of UMBRA for each Spot (and for Day)	Area of WHOLE for each Spot (and for Day)	Area for each Group (and for Day)
1881.								
137d.941	478	0·460	328·0	176·6	+21·0	1	5	
		0·688	131·5					182
		0·858	113·7					149
		0·884	58·0					245
May 19						(60)	(226)	(1551)
138·931	.	0·960	243·8					143
	472	0·870	287·9	207·0	+14·6	18	164	618 n
	477	0·625	221·3	176·6	−29·5	31	92	
	477	0·600	212·6	170·7	−31·9	6	50	
	478	0·583	311·1	176·3	+20·9	0	6	
May 20						(65)	(312)	(1258)
139·917		0·927	244·7					79
	472	0·960	285·6	208·9	+14·5	7	98	1213
	477	0·765	231·2	179·5	−29·7	13	69	
	477	0·706	222·8	170·8	−32·4	4	17	
	479	0·531	240·9	165·2	−16·3	0	7	
	480	0·561	221·0	160·5	−26·4	30	95	
	480	0·544	213·6	156·3	−28·3	18	53	
		0·941	65·0					238
May 21						(72)	(339)	(1530)
140·691		0·974	286·1					392
		0·912	293·6					913
	477	0·851	235·1	178·8	−30·1	19	47	
	480	0·661	230·0	159·8	−26·0	22	111	
	480	0·616	222·6	154·0	−28·8	21	40	
	481	0·679	131·2	90·0	−27·9	4	20	
	481	0·712	129·5	86·8	−28·1	2	28	
May 22*						(66)	(246)	(1305)
142·063		0·935	294·0					750
		0·822	283·2					683
	477	0·955	239·4	177·6	−29·6	16	58	
	477	0·919	236·2	170·2	−31·4	17	44	581 c
	480	0·833	240·6	160·1	−25·0	23	60	377 n
	480	0·787	233·6	153·1	−28·8	2	9	179 p
	481	0·503	151·5	91·4	−27·5	23	97	
	481	0·561	144·7	85·4	−28·5	20	60	
	482	0·839	66·3	50·7	+19·4	54	99	
	482	0·904	67·2	44·9	+19·8	3	19	190 c
		0·836	131·2					101
May 23						(158)	(446)	(2861)
142·886		0·978	294·3					432
		0·948	280·3					318
		0·825	284·7					323
	477	0·968	237·6	169·5	−31·6	0	32	493 c
	480	0·916	243·6	160·3	−24·6	15	48	35g n
	480	0·871	237·3	132·6	−28·8	0	9	251 c
	481	0·450	172·0	92·0	−27·7	22	103	
	481	0·439	161·0	85·7	−28·8	11	38	
	482	0·748	61·6	51·9	+19·8	36	113	
	482	0·772	63·4	49·3	+19·2	0	15	
1881.								
142d·886	482	0·810	65·2	45·3	+19·0	0	27	191 c
		0·882	134·5					176
May 24						(84)	(385)	(2543)
143·896		0·925	281·5					33g
		0·872	297·5					346
	480	0·982	245·2	161·1	−24·6	0	35	331 sf
	481	0·471	197·2	91·8	−27·8	42	219	
	481	0·453	189·7	87·7	−27·6	13	23	
	481	0·471	184·9	85·4	−29·1	7	16	
	482	0·590	52·5	53·0	+19·9	30	109	
	482	0·636	55·7	48·9	+20·0	9	20	
	482	0·682	58·8	44·6	+19·6	36	79	
May 25						(137)	(501)	(1016)
144·783		0·968	282·3					322
	481	0·546	214·6	91·4	−27·7	43	196	
	481	0·517	207·2	86·5	−28·3	4	34	
	482	0·460	38·6	53·3	+20·0	22	100	
	482	0·523	46·2	47·4	+20·2	15	67	
		0·974	112·3					676
May 26*						(84)	(397)	(998)
145·784	481	0·672	227·7	91·7	−27·7	33	205	
	481	0·632	224·5	87·6	−27·6	11	91	
	482	0·365	111·7	53·2	+19·9	18	71	
	482	0·380	22·2	49·0	+19·5	4	44	
	482	0·437	27·5	45·2	+21·7	13	73	
	485	0·947	99·7	346·8	− 9·5	0	71	
	485	0·963	100·1	343·7	−10·0	30	144	1063 c
	485	0·983	101·3	338·5	−11·3	52	331	
		0·936	112·3					1012
May 27*						(161)	(1030)	(2075)
146·692	481	0·776	234·6	90·9	−27·3	47	246	
	481	0·742	231·3	86·7	−28·2	0	68	
	482	0·375	340·5	53·3	+19·7	16	53	
	482	0·356	350·1	49·4	+19·9	5	49	
	482	0·374	2·9	44·6	+20·9	0	50	
	484	0·819	112·6	351·8	−19·1	0	32	
	485	0·860	99·7	346·9	− 8·8	35	105	
	485	0·886	100·2	343·8	− 9·4	3g	151	1589 c
	485	0·925	101·2	338·5	−10·7	150	355	
		0·888	116·8					1027
May 28*						(292)	(1109)	(2616)
147·601	481	0·874	239·3	90·9	−26·9	47	215	
	481	0·839	237·5	86·3	−27·3	0	51	
	482	0·474	31·8	53·3	+19·8	15	54	
	482	0·444	322·5	50·3	+19·8	11	49	
	482	0·413	334·1	44·8	+20·9	0	25	
	484	0·692	115·2	352·8	−17·7	0	5	
	484	0·705	116·7	352·1	−19·0	0	21	
	485	0·729	101·4	347·6	− 8·9	17	79	
	485	0·769	100·5	344·1	− 9·3	26	161	739 c
	485	0·826	102·6	338·7	−10·9	130	376	
May 29*						(246)	(1046)	(739)

* Indian photo.   The Areas of Spots and Faculæ are expressed in Millionths of the Sun's visible Hemisphere.

MEASURES of POSITIONS and AREAS of SPOTS and FACULÆ upon the SUN's DISK on PHOTOGRAPHS taken in the YEAR 1881—*continued.*

Mean Solar Time	No. of Group, and Letter for Spot	Distance from Centre in terms of Sun's Radius	Position Angle from Sun's Axis	Longitude	Latitude	Area of UMBRA for each Spot (and for Day)	Area of WHOLE for each Spot (and for Day)	Area for each Group (and for Day)
1881.								
148·992	481	0·976	242·6	91·3	−26·8	83	318	428 s f
	482	0·670	301·8	52·5	+20·1	14	90	167 c
	483	0·617	333·5	34·4	+32·8	0	16	
	483	0·611	328·2	37·2	+30·6	0	11	
	484	0·492	130·2	352·0	−19·0	0	6	
	485	0·478	108·0	348·0	− 9·0	11	81	
	485	0·529	107·5	344·7	− 9·7	19	142	
	485	0·615	107·8	338·8	−11·3	116	547	715 c
		0·703	131·1					154
		0·824	121·7					155
		0·928	84·0					57
		0·963	69·7					158
May 30						(243)	(1211)	(1834)
149·896		0·936	299·8					151
		0·704	319·8					103
	482	0·795	296·0	52·7	+20·0	11	42	472 n f
	485	0·297	118·7	348·2	− 8·6	14	57	
	485	0·349	116·8	345·0	− 9·5	20	126	
	485	0·450	113·6	338·6	−10·8	128	702	
	486	0·711	127·0	324·5	−25·7	0	9	108 c
	487	0·970	66·3	289·1	+22·8	9	33	474 s p
		0·850	83·2					73
		0·866	103·8					73
May 31						(182)	(969)	(1454)
150·918		0·864	294·5					592
	485	0·121	183·1	350·2	− 7·3	0	4	
	485	0·145	170·9	348·5	− 8·6	0	11	
	485	0·167	165·1	347·3	− 9·6	3	40	
	485	0·176	156·3	345·7	− 9·6	25	124	
	485	0·252	133·6	339·2	−10·3	158	931	
	487	0·847	63·4	295·2	+21·0	10	29	530 c
	487	0·898	63·7	288·9	+23·3	11	57	110
		0·914	118·7					
June 1						(207)	(1196)	(1232)
151·904								576
	485	0·946	292·8					
	485	0·254	234·2	348·7	− 8·8	0	4	
	485	0·235	225·6	346·5	− 9·7	18	111	
	485	0·198	207·8	342·1	−10·3	34	239	
	485	0·167	189·1	338·3	− 9·7	94	466	
	487	0·723	59·3	295·0	+21·4	26	93	
	487	0·767	61·3	290·7	+21·3	34	89	184 f
	487	0·797	60·2	288·2	+23·1	0	24	216
		0·923	71·1					220
		0·930	110·8					
June 2						(206)	(1026)	(1196)
152·938	485	0·436	248·6	347·3	− 9·3	24	87	
	485	0·370	246·4	343·0	− 8·6	14	102	
	485	0·375	241·0	342·5	−10·6	23	69	
	485	0·359	236·3	340·7	−11·6	0	19	
	485	0·330	246·2	340·7	− 7·8	5	28	
	485	0·339	240·0	340·3	− 9·9	11	51	
1881.								
	485	0·351	231·4	339·5	−12·7	0	14	
	485	0·298	140·5	338·2	− 8·5	78	331	
	487	0·571	50·1	295·1	+21·3	24	88	
	487	0·625	54·6	290·1	+21·1	28	155	
		0·838	66·3					134
		0·840	112·7					312
June 3						(207)	(944)	(446)
153·901								340
	485	0·736	243·7					
	485	0·621	254·7	347·6	− 9·4	19	106	
	485	0·569	254·9	343·9	− 8·5	13	32	
	485	0·569	251·6	343·5	−10·3	12	26	
	485	0·532	251·4	341·0	− 9·8	11	56	
	485	0·489	253·5	338·5	− 8·0	47	301	499 c
	487	0·430	34·9	295·2	+20·5	23	124	
	487	0·487	41·8	290·1	+21·2	31	203	
		0·738	118·8					283
June 4						(156)	(848)	(1122)
154·594		0·789	246·1					699
	485	0·738	256·9	347·8	− 9·5	24	78	545 c
	485	0·694	258·1	344·3	− 8·1	0	30	
	485	0·621	256·9	338·7	− 8·0	53	154	
	485	0·617	253·0	337·9	−10·3	0	40	
	487	0·367	12·2	295·5	+21·1	12	48	
	487	0·353	14·1	296·0	+20·0	7	27	
	487	0·361	16·0	295·1	+20·3	7	27	
	487	0·466	27·2	289·8	+21·2	27	116	
		0·838	7·2					604
June 5†						(130)	(520)	(1848)
156·040		0·877	238·5					298
		0·410	297·3					83
	485	0·918	159·5	348·0	− 9·5	12	26	
	485	0·898	261·1	345·4	− 7·9	0	10	
	485	0·845	260·6	339·1	− 7·8	22	64	1185 c
	487	0·426	326·2	296·7	+20·9	9	25	
	487	0·384	340·4	290·0	+21·4	22	79	
	487	0·391	3·8	280·4	+23·1	0	6	
	487	0·406	9·6	277·8	+23·7	0	3	
		0·368	122·5					348
June 6						(65)	(213)	(1914)
158·032		0·990	259·5					326
		0·953	297·2					280
		0·933	241·0					313
	487	0·630	304·4	289·4	+21·2	7	23	445 p
	487	0·555	313·2	281·6	+22·7	11	19	
	487	0·558	317·6	280·0	+24·7	0	8	
	487	0·544	322·0	277·4	+25·7	4	29	
	488	0·438	51·1	334·9	+16·3	2	15	
	488	0·491	53·1	231·4	+17·5	4	16	
		0·873	68·7					251
		0·988	119·0					517
June 8						(28)	(110)	(2132)

The Areas of Spots and Faculæ are expressed in Millionths of the Sun's visible Hemisphere.   † Melbourne photo.

MEASURES of POSITIONS and AREAS of SPOTS and FACULÆ upon the SUN'S DISK on PHOTOGRAPHS taken in the YEAR 1881—*continued.*

Mean Solar Time	No. of Group, and Letter for Spot	Distance from Centre in terms of Sun's Radius	Position Angle from Sun's Axis	Heliographic Longitude	Heliographic Latitude	Area of UMBRA for each Spot (and for Day)	Area of WHOLE Spot (and for Day)	Area for each Group (and for Day)
1881.			°	°	°			
158ᵈ·698	487	0·726	299·2	289·4	+21·1	0	29	
	487	0·657	305·5	282·1	+22·8	11	26	
	487	0·648	308·6	280·4	+24·3	6	22	
	487	0·619	313·3	276·6	+25·5	11	40	
	488	0·396	40·7	231·1	+18·0	0	18	
	489	0·924	74·5	180·1	+14·5	3g	270	1008 f
		0·848	67·9					342
		0·976	121·0					1346
June 9*						(67)	(405)	(2696)
156·937	487	0·829	297·9	283·1	+23·2	15	39	
	487	0·810	299·8	280·6	+24·2	0	23	1123 s p
	489	0·792	71·7	179·6	+14·8	82	306	
	489	0·821	72·3	176·7	+14·9	0	26	1258 f
	490	0·863	125·1	176·6	−29·3	15	36	
	490	0·920	122·7	167·9	−29·5	0	29	1304 c
	491	0·931	100·5	162·5	− 9·5	14	67	601 f
June 10						(126)	(526)	(4286)
160·911		0·865	244·8					124
	487	0·928	294·8	283·8	+23·2	0	10	738 c
	489	0·647	67·8	179·4	+14·8	30	244	
	489	0·680	68·7	176·8	+14·9	9	110	1449 f
	490	0·761	130·9	176·5	−29·2	15	40	
	490	0·775	128·2	174·1	−28·0	0	5	
	490	0·838	126·9	167·3	−29·6	0	30	979 f
	491	0·838	102·7	161·6	−10·1	12	109	
	491	0·863	99·9	158·6	− 8·1	0	12	
	491	0·885	101·4	156·2	− 9·6	9	78	807 c
June 11						(95)	(638)	(4097)
162·917		0·914	245·2					173
		0·819	233·8					191
	492	0·433	305·6	212·4	+15·5	0	4	
	492	0·396	311·9	208·8	+16·3	0	4	
	489	0·313	39·5	179·2	+15·0	49	195	
	489	0·298	45·9	178·4	+13·0	0	8	
	489	0·338	41·7	177·6	+15·6	0	22	
	489	0·339	45·9	176·5	+14·6	9	106	
	493	0·370	5·5	168·7	+22·6	6	15	
	490	0·555	155·1	175·5	−29·1	5	29	
	490	0·571	152·2	173·3	−29·2	0	26	
	490	0·621	144·2	166·5	−29·2	0	31	
	491	0·515	110·3	161·8	− 9·3	9	72	
	491	0·595	107·4	156·0	− 9·3	8	40	
	494	0·952	113·8	121·0	−22·2	40	172	426 s
		0·756	138·3					667
		0·880	111·0					149
June 13						(126)	(724)	(1606)
163·901		0·844	289·0					140
		0·795	234·7					347
	492	0·590	296·7	211·2	+16·3	0	14	
	492	0·561	297·9	208·9	−16·2	5	9	
	489	0·245	355·6	179·1	+15·3	33	197	
	489	0·240	5·0	176·7	+15·0	6	73	
	490	0·506	175·2	175·2	−29·0	0	25	
1881.			°	°	°			
163ᵈ·901	490	0·506	172·6	173·7	−28·8	0	11	
	491	0·308	125·9	163·4	− 9·2	14	28	
	491	0·33g	126·1	161·9	−10·3	0	5	
	491	0·405	116·8	156·6	− 9·4	0	7	
	495	0·680	118·6	139·2	−18·0	9	19	
	495	0·712	118·1	136·6	−18·7	0	22	78 c
	494	0·867	116·2	121·3	−21·8	16	108	491 f
June 14						(83)	(518)	(1056)
164·711		0·921	288·5					360
		0·870	235·1					734
	492	0·732	290·9	212·4	+16·0	0	44	
	492	0·686	293·9	208·0	+17·1	9	30	
	489	0·316	320·6	179·2	+15·4	30	171	
	489	0·282	326·0	176·5	+14·8	2	60	
	490	0·508	193·7	175·0	−28·2	0	44	190 c
	491	0·194	161·9	163·7	− 9·3	3	15	
	491	0·210	151·4	161·4	− 9·3	0	24	
	491	0·248	141·1	158·1	− 9·8	0	15	
	491	0·257	133·4	156·4	− 8·9	0	22	
	495	0·539	128·3	140·9	−18·4	20	39	
	495	0·602	124·5	135·7	−18·9	21	50	
	494	0·776	120·4	121·1	−22·2	29	105	
	495a	0·984	85·9	87·7	+ 4·3	27	73	
	496	0·995	117·2	84·2	−27·0	0	159	
June 15*						(141)	(851)	(1284)
165·705	489	0·479	300·9	179·2	+15·5	25	176	
	495	0·405	150·2	141·8	−19·2	18	72	
	495	0·459	139·3	135·7	−18·9	19	75	
	494	0·643	128·2	121·2	−22·2	15	97	
	496	0·956	119·1	84·4	−27·2	3g	134	560 c
June 16*						(116)	(554)	(560)
166·747	489	0·956	286·3					756
	489	0·660	292·3	179·5	+15·7	31	187	
	495	0·360	186·6	142·8	−19·4	18	53	
	495	0·361	167·0	135·4	−19·0	18	53	
	494	0·505	143·0	121·2	−22·3	23	73	
	496	0·871	122·4	84·9	−26·9	17	151	
		0·898	125·7					560
June 17*						(107)	(517)	(560)
167·691		0·916	291·7					783
		0·808	231·9					490
	489	0·802	288·7	179·8	+15·9	33	132	276 c
	495	0·442	215·5	143·5	−19·5	20	70	
	495	0·370	198·3	134·8	−18·9	0	16	
	494	0·420	165·4	121·3	−22·4	22	78	
	496	0·770	127·7	84·9	−26·8	28	129	
	496	0·807	127·8	81·5	−28·4	6	31	
		0·843	133·1					567
June 18*						(109)	(456)	(2116)

* Indian photo.　　The Areas of Spots and Faculæ are expressed in Millionths of the Sun's visible Hemisphere.

MEASURES of POSITIONS and AREAS of SPOTS and FACULÆ upon the SUN's DISK on PHOTOGRAPHS taken in the YEAR 1881—*continued.*

Mean Solar Time.	No. of Group, and Letter for Spot.	Distance from Centre in terms of Sun's Radius.	Position Angle from Sun's Axis.	HELIOGRAPHIC Longitude.	HELIOGRAPHIC Latitude.	SPOTS Area of UMBRA for each Spot (and for Day).	SPOTS Area of WHOLE for each Spot (and for Day).	FACULÆ Area for each Group (and for Day).
1881. 169d.904		0.894	256.5					344
		0.872	233.0					739
		0.854	288.2					363
	495	0.766	242.3	144.3	-19.5	0	27	
	494	0.546	220.6	121.0	-22.7	9	28	
	496	0.524	154.8	84.1	-26.4	18	83	
	497	0.525	61.4	70.0	+16.2	0	8	
		0.863	62.8					229
June 20						(27)	(146)	(1675)
170.982		0.976	258.0					495
		0.938	234.5					750
		0.925	183.3					493
	495	0.892	247.4	144.6	-19.0	13	19	256 f
	494	0.684	234.0	120.8	-22.0	2	12	
	496	0.476	179.7	84.1	-26.3	13	47	
June 21						(28)	(78)	(1994)
171.938		0.942	287.3					146
		0.861	244.0					165
	496	0.513	200.7	83.2	-26.5	7	32	
		0.962	106.3					569
June 22						(7)	(32)	(880)
172.961		0.896	241.3					241
	495	0.993	250.8	139.8	-18.8	0	21	481 sf
	498	0.761	288.1	106.4	+15.1	0	6	
	498	0.715	287.6	102.5	+14.1	7	18	
	496	0.610	218.3	83.0	-26.5	13	22	
	499	0.455	144.7	42.0	-19.6	0	6	
	500	0.918	100.4	352.3	-8.6	4	33	
	500	0.980	103.5	340.7	-12.8	4	17	1372 c
	501	0.998	64.3	331.3	+25.8	108	515	182 p
		0.915	68.7					64
June 23						(136)	(638)	(2340)
173.975		0.967	232.5					267
	498	0.890	286.3	106.9	+15.6	16	60	
	498	0.867	287.2	103.9	+16.1	94	154	
	498	0.855	284.8	102.8	+13.9	15	98	398 c
	496	0.734	229.0	82.9	-26.9	9	28	452 s
	500	0.789	103.0	353.8	-8.7	0	20	
	500	0.816	196.1	351.7	-11.7	0	25	871 n
	501	0.962	63.5	331.5	+26.1	73	381	
	502	0.994	61.6	320.8	+28.5	0	110	690 n
June 24						(207)	(876)	(2678)
174.712	498	0.954	285.3	107.2	+15.3	27	203	
	498	0.929	285.9	102.8	+15.7	0	44	
	498	0.930	283.3	103.1	+13.5	45	121	739 c
	496	0.815	233.5	82.1	-27.1	0	40	783 c
	500	0.672	105.6	354.2	-8.6	2	18	

Mean Solar Time.	No. of Group, and Letter for Spot.	Distance from Centre in terms of Sun's Radius.	Position Angle from Sun's Axis.	HELIOGRAPHIC Longitude.	HELIOGRAPHIC Latitude.	SPOTS Area of UMBRA for each Spot (and for Day).	SPOTS Area of WHOLE for each Spot (and for Day).	FACULÆ Area for each Group (and for Day).
1881. 174d.712	501	0.902	61.3	332.7	+26.8	42	133	
	501	0.913	62.9	330.8	+25.6	12	73	1277 c
	502	0.964	60.7	321.5	+28.8	0	86	
		0.969	12.6					891
		0.988	76.9					624
June 25*						(164)	(725)	(3252)
176.780		0.973	240.5					850
	503	0.321	158.8	0.7	-14.7	31	122	
	503	0.355	150.1	357.1	-15.2	4	35	
	503	0.368	144.9	355.0	-14.8	20	98	
	501	0.659	51.5	332.6	+26.3	29	143	
	501	0.674	54.7	330.5	+25.0	22	74	
	502	0.788	55.4	320.3	+28.4	13	54	
	504	0.755	73.3	319.3	+12.8	25	126	
	504	0.787	76.6	316.2	+12.2	11	67	
	505	0.847	77.3	310.1	+12.2	6	78	353 c
	505	0.858	74.0	309.2	+15.1	19	176	
		0.862	136.8					325
		0.983	73.5					1061
June 27*						(180)	(973)	(2589)
178.027		0.802	299.7					230
	503	0.331	207.3	359.9	-14.3	64	250	
	503	0.309	192.4	354.9	-14.7	48	251	
	503	0.303	362.2	331.7	+26.5	31	144	
	501	0.503	47.0	329.4	+24.5	10	65	
	502	0.639	47.2	319.0	+28.0	11	39	571 nf
	504	0.538	70.6	319.8	+12.7	34	171	
	504	0.574	72.9	317.0	+12.0	5	34	170 c
	505	0.684	73.7	308.8	+13.1	9	70	353 c
	505	0.674	70.6	310.1	+15.0	10	80	
	505	0.727	70.2	305.7	+16.1	13	111	
	506	0.926	68.2	284.0	+21.2	0	17	
	506	0.928	66.2	283.9	+23.1	30	121	
	506	0.958	65.2	278.3	+24.6	0	43	1277 c
		0.747	145.7					394
June 28						(265)	(1396)	(2995)
178.916		0.882	297.5					245
	503	0.465	229.1	0.5	-14.9	69	270	
	503	0.404	221.5	355.2	-14.8	65	218	
	503	0.350	226.5	354.2	-11.1	7	30	
	501	0.425	15.9	332.2	+26.9	32	135	
	501	0.417	24.1	328.4	+25.1	10	50	
	504	0.363	62.1	320.2	+12.5	29	118	
	502	0.508	34.1	320.6	+27.6	8	23	
	505	0.506	65.8	310.9	+14.5	17	72	
	505	0.535	69.9	308.3	+13.1	13	49	
	505	0.579	65.7	306.1	+16.2	21	166	
	506	0.836	66.4	284.2	+21.2	0	9	
	506	0.845	64.1	283.6	+23.3	38	142	
	506	0.851	68.1	282.2	+20.1	0	10	
	506	0.895	63.3	277.5	+25.1	0	67	1059 c
June 29						(309)	(1359)	(1304)

* Indian photo.   The Areas of Spots and Faculæ are expressed in Millionths of the Sun's visible Hemisphere.

MEASURES of POSITIONS and AREAS of SPOTS and FACULÆ upon the SUN'S DISK on PHOTOGRAPHS taken in the YEAR 1881—*continued.*

Mean Solar Time.	No. of Group, and Letter for Spot.	Distance from Centre in terms of Sun's Radius.	Position Angle from Sun's Axis.	HELIOGRAPHIC Longitude.	HELIOGRAPHIC Latitude.	Area of UMBRA for each Spot (and for Day).	Area of WHOLE Spot (and for Day).	Area for each Group (and for Day).
1881. 179ᵈ·981		°	°	°	°			
	503	0·961	295·5	359·9	−14·4	45	238	162
	503	0·626	242·2	355·1	−14·5	41	206	
	501	0·423	346·8	331·3	+27·2	25	120	
	504	0·187	30·3	319·6	+12·3	10	33	
	502	0·445	13·4	318·4	+28·6	0	14	
	505	0·316	49·3	310·9	+14·8	13	54	
	505	0·301	56·1	310·4	+12·6	4	27	
	503	0·400	53·9	305·5	+16·4	24	113	
	306	0·657	64·8	286·5	+18·6	0	12	
	506	0·691	61·5	284·6	+21·5	2	8	
	506	0·715	59·3	283·2	+23·6	21	95	
	506	0·733	60·8	281·3	+23·1	0	5	1003 c
		0·812	122·5					394
June 30						(185)	(925)	(1559)
181·022	503	0·780	148·4	359·7	−14·6	78	289	
	503	0·751	245·4	356·4	−16·0	13	39	
	503	0·733	248·8	355·7	−13·1	14	77	
	503	0·728	246·7	354·8	−14·4	23	81	
	503	0·706	247·3	353·2	−13·4	9	23	779 c
	501	0·515	323·8	331·3	+27·4	32	117	
	501	0·463	326·0	327·9	+25·5	1	37	
	502	0·421	345·8	318·0	+27·1	8	18	
	504	0·207	321·6	318·9	+12·4	6	27	
	505	0·205	1·1	311·1	+14·9	18	75	
	505	0·247	20·6	306·2	+16·4	5	37	
	506	0·531	51·7	284·8	+22·0	3	8	
	506	0·569	50·7	282·7	+23·8	24	86	
	507	0·979	67·2	233·0	+23·0	0	135	305 n p
		0·725	52·7					209
		0·795	120·8					103
		0·931	105·0					176
July 1						(234)	(1049)	(1572)
181·919	503	0·891	251·7	0·0	−14·6	25	297	
	503	0·856	252·1	355·9	−13·5	0	85	
	503	0·846	250·4	354·5	−14·6	18	101	
	503	0·831	251·6	353·2	−13·3	0	15	1265
	501	0·627	312·3	330·7	+27·7	25	70	
	502	0·482	323·8	317·8	+25·7	3	10	
	504	0·363	297·6	318·4	+12·7	0	38	
	505	0·287	314·9	311·4	+14·8	15	87	
	505	0·259	336·8	306·8	+16·2	3	33	
	506	0·449	37·3	282·1	+23·9	17	72	
	507	0·911	66·9	233·0	+22·5	22	134	338 s
		0·701						166
		0·981						1497
July 2						(128)	(942)	(3266)
182·615	503	0·945	252·4	359·2	−15·4	50	215	
	503	0·921	263·5	355·5	−13·8	38	170	966 c
	503	0·904	252·0	352·8	−14·7	23	127	
	501	0·722	305·4	331·5	+27·1	21	84	
	504	0·494	291·6	318·4	+13·4	7	27	

Mean Solar Time.	No. of Group, and Letter for Spot.	Distance from Centre in terms of Sun's Radius.	Position Angle from Sun's Axis.	HELIOGRAPHIC Longitude.	HELIOGRAPHIC Latitude.	Area of UMBRA for each Spot (and for Day).	Area of WHOLE Spot (and for Day).	Area for each Group (and for Day).
1881. 182ᵈ·615	505	0·415	300·1	312·0	+15·0	18	55	
	506	0·374	18·8	282·8	+23·9	20	98	
	507	0·844	64·9	234·5	+22·8	37	123	447 c
		0·970	130·2					668
		0·972	71·9					1903
July 3*						(209)	(899)	(3984)
183·944								842
								234
	501	0·974	255·3					
	501	0·841	222·5					
	501	0·869	300·5	330·4	+27·9	31	80	
	501	0·859	297·9	329·7	+25·6	3	15	1076 n f
	504	0·726	285·1	318·5	+13·3	0	8	231
	505	0·646	289·6	311·6	+15·2	10	19	
	506	0·364	333·2	282·8	+21·3	0	19	
	506	0·383	338·1	281·6	+24·1	15	61	
	506	0·411	352·8	276·0	+27·4	0	2	
	507	0·690	60·3	232·4	+22·6	14	80	
	508	0·923	65·3	206·2	+24·3	47	95	
	509	0·943	69·3	202·4	+20·7	0	27	929 s p
	510	0·966	111·5	200·2	−19·7	0	8	116 f
	515	0·960	74·3	198·8	+16·0	9	47	
	511	0·983	73·7	192·9	+16·7	0	25	540 c
		0·866	125·7					201
		0·978	117·8					81
July 4						(129)	(486)	(4250)
184·960								250
	501	0·912	229·5					354
		0·865	283·3					1166 n f
	501	0·950	297·8	330·3	+27·5	19	63	
	506	0·496	315·6	281·5	+24·0	13	61	
	512	0·393	143·7	245·4	−14·9	0	4	
	512	0·413	139·3	243·1	−14·7	0	1	
	513	0·427	136·6	241·7	−14·6	4	7	
	507	0·542	50·9	232·1	+23·1	0	32	
	508	0·806	63·7	207·7	+23·1	0	54	492 c
	509	0·849	68·0	202·3	+20·5	0	23	89 f
	510	0·884	114·2	200·7	−19·4	0	15	250 f
	515	0·861	73·6	200·1	+15·9	11	124	154 c
	511	0·918	73·2	192·8	+16·9	25	82	504 f
	513	0·906	120·6	199·5	−25·7	8	21	92 c
	513	0·929	119·9	195·8	−26·0	0	12	
July 5						(100)	(499)	(3451)
185·949								1123
	506	0·918	285·5					
	506	0·638	304·5	281·2	+24·2	11	43	
	508	0·643	57·1	210·2	+23·5	0	15	
	508	0·657	58·5	208·8	+23·0	0	4	
	515	0·705	71·8	201·3	+15·4	23	77	
	515	0·711	70·9	201·1	+16·3	26	91	
	509	0·721	61·5	202·2	+20·7	5	24	
	511	0·814	71·5	192·4	+17·2	37	122	
		0·895	72·5					995
		0·982	103·3					813
July 6*						(102)	(376)	(2931)

* Indian photo.   The Areas of Spots and Faculæ are expressed in Millionths of the Sun's visible Hemisphere.

MEASURES of POSITIONS and AREAS of SPOTS and FACULÆ upon the SUN's DISK on PHOTOGRAPHS taken in the YEAR 1881—*continued.*

Mean Solar Time	No. of Group and Letter for Spot	Distance from Centre in terms of Sun's Radius	Position Angle from Sun's Axis	Heliographic Longitude	Heliographic Latitude	Area of UMBRA for each Spot (and for Day)	Area of WHOLE Spot for each (and for Day)	Faculæ Area for each Group (and for Day)
1881.								
186d·608	506	0·733	300·1	281·3	+24·2	0	29	576 s p
	508	0·451	56·0	214·3	18·0	0	11	
	508	0·596	50·5	206·9	25·4	0	21	
	508	0·552	50·8	209·7	23·7	0	14	
	514	0·201	0·4	237·3	+15·2	15	39	
	515	0·592	69·3	202·5	+15·1	33	160	
	509	0·627	61·0	201·7	+20·6	6	28	
	515a	0·647	67·4	198·9	+17·2	0	46	
	511	0·726	70·1	192·1	+16·8	22	131	
		0·743	70·0					684
		0·939	114·1					1322
July 7†						(76)	(479)	(2582)
188·926		0·954	290·8					270
		0·888	245·0					275
		0·885	302·3					299
	514	0·536	293·1	237·4	+15·5	11	40	
	514	0·502	296·7	234·5	+16·5	3	5	
	508	0·335	351·5	209·8	+23·2	6	27	
	508	0·381	3·6	205·2	+26·2	0	4	
	509	0·310	16·9	201·2	+21·1	0	8	
	515	0·176	14·6	204·1	+13·7	20	165	
	515	0·203	15·3	203·6	+15·2	0	4	
	511	0·325	46·8	192·5	+16·7	12	75	
	516	0·916	114·2	144·0	−20·2	17	122	
	516	0·927	112·3	141·9	−18·9	62	299	
	516	0·936	115·3	141·1	−22·0	0	11	634 f
		0·835	133·7					187
July 9						(131)	(760)	(1665)
191·027		0·928	253·5					91
		0·771	230·8					294
	514	0·872	284·9	239·4	+15·0	4	18	386 f
	515	0·464	292·6	205·0	+14·0	57	269	
	515	0·411	301·7	200·2	+16·3	12	127	
	515	0·681	121·6	196·5	+16·7	9	58	
	516	0·365	307·3	143·5	−20·7	16	106	
	516	0·689	124·2	141·9	−19·4	68	303	1841 f
	517	0·909	75·7	113·5	+13·9	25	125	452 f
July 11						(191)	(1006)	(3064)
191·990		0·849	292·2					666
	515	0·645	287·1	205·5	+14·2	62	281	
	515	0·593	290·9	201·2	+15·7	11	46	
	511	0·639	293·0	196·7	+16·8	6	33	
	516	0·554	136·5	142·4	−19·7	74	393	
	518	0·773	113·1	119·0	−14·7	0	41	
	518	0·798	112·4	116·6	−14·9	0	7	121 c
	517	0·794	75·8	113·9	+13·8	20	101	555 f
		0·866	141·5					583
July 12						(173)	(902)	(1926)
192·985		0·866	238·7					463
	515	0·797	284·6	205·5	+14·2	42	272	
	515	0·757	287·2	201·6	+15·8	5	45	1124 н
	519	0·754	240·7	196·7	−18·4	0	5	

Mean Solar Time	No. of Group and Letter for Spot	Distance from Centre in terms of Sun's Radius	Position Angle from Sun's Axis	Heliographic Longitude	Heliographic Latitude	Area of UMBRA for each Spot (and for Day)	Area of WHOLE Spot for each (and for Day)	Faculæ Area for each Group (and for Day)
1881.								
192d·985	516	0·444	156·7	142·3	−19·7	71	407	
	520	0·532	147·3	135·0	−22·3	0	13	
	518	0·610	120·3	120·2	−14·2	7	32	
	517	0·638	72·9	114·2	+14·2	19	77	380 f
		0·749	156·7					224
July 13						(144)	(851)	(2191)
193·978		0·896	240·5					415
	515	0·913	283·8	205·9	+14·4	71	256	
	515	0·885	285·3	202·1	+15·6	0	29	1971 f
	516	0·417	191·7	145·1	−19·6	0	5	
	516	0·414	185·1	142·2	−19·8	52	418	
	520	0·459	169·2	134·6	−22·2	1	17	
	518	0·442	134·7	121·1	−13·8	4	15	
	517	0·455	66·9	114·4	+14·3	24	83	
		0·914	61·0					310
		0·915	123·3					201
July 14						(152)	(823)	(2897)
194·900		0·951	244·8					480
	515	0·978	283·4	206·1	+14·0	37	264	780 н J
	516	0·453	210·2	142·3	−19·6	73	425	
	520	0·489	193·7	134·6	−22·4	0	63	
	518	0·339	160·3	121·0	−13·9	6	23	
	517	0·276	51·5	114·9	+14·3	16	33	
		0·789	57·5					231
		0·884	128·0					509
		0·922	110·7					198
July 15						(132)	(808)	(2198)
195·895		0·980	250·3					201
		0·918	286·0					653
		0·872	232·3					245
		0·850	250·8					533
	516	0·618	125·6	142·7	−21·5	18	55	
	516	0·593	228·5	142·4	−18·9	65	337	
	518	0·338	199·3	121·1	−13·9	0	10	
	518	0·329	193·6	119·1	−13·9	0	4	
	517	0·174	357·0	115·1	+14·6	4	14	
		0·834	134·5					268
July 16						(87)	(420)	(1900)
196·641	516	0·719	233·4	142·9	−21·7	12	33	
	516	0·702	238·0	143·2	−18·1	7	30	
	516	0·693	235·9	141·7	−19·0	32	121	
	517	0·260	312·9	115·9	+14·8	2	9	
July 17*						(53)	(193)	
197·981		0·807	225·3					692
	516	0·871	242·3	142·5	−21·2	0	17	
	516	0·857	242·1	141·4	−19·2	17	59	370 c
	518	0·638	240·3	121·7	−14·5	11	23	
	518	0·618	238·6	119·8	−14·7	0	10	
	521	0·826	76·1	31·4	+14·2	26	122	
	521	0·880	76·8	25·2	+13·9	0	52	323 c
	522	0·973	72·2	9·6	+18·5	115	382	390 f
July 18						(169)	(665)	(1775)

* Indian photo.   The Areas of Spots and Faculæ are expressed in Millionths of the Sun's visible Hemisphere.   † Melbourne photo.

as of Spots and Faculæ upon the Sun's Disc on Photographs taken in the Year 1881—*continued.*

Heliographic Latitude	Spots — Area of UMBRA for each Spot (and for Day)	Spots — Area of WHOLE Spot for each Spot (and for Day)	Faculæ — Area for each Group (and for Day)	Mean Solar Time	No. of Group and Letter for Spot	Distance from Centre in terms of Sun's Radius	Position Angle from Sun's Axis	Heliographic Longitude	Heliographic Latitude	Spots — Area of UMBRA for each Spot (and for Day)	Spots — Area of WHOLE Spot for each Spot (and for Day)	Faculæ — Area for each Group (and for Day)
			413	1881. 203ᵈ·920	525	0·830	63·2	313·7	+25·1	9	44	
			160		526	0·866	62·4	309·6	+26·5	36	313	395 c
−19·3	21	97	567 a f		527	0·847	109·7	313·5	−13·6	0	34	
−13·9	12	31			527	0·872	109·3	310·6	−13·9	0	30	
−13·3	0	8	189 c		528	0·918	108·1	304·4	−14·3	12	83	
+14·5	0	11				0·959	108·9					893
+14·6	32	146		July 24*						(209)	(1453)	(1603)
+13·5	12	34										
+13·1	12	48										
+14·5	0	26	222 c	204·966	524	0·939	232·0					136
+18·7	86	347	348 f		524	0·854	285·6	53·2	+16·2	29	96	
(175)	(748)		(1899)		524	0·838	284·5	51·5	+15·1	0	14	
			695		524	0·816	287·4	48·9	+17·3	13	53	
			504		524	0·807	289·1	47·9	+18·6	4	11	243 c
+15·1	24	134			521	0·618	278·2	32·7	+ 9·3	15	23	
+13·9	8	43			521	0·600	279·9	31·3	+10·3	3	16	
+13·7	17	51			521	0·597	277·5	31·2	+ 8·8	0	4	
+19·3	85	331			521	0·602	286·8	30·9	+14·4	15	90	
			1000		521	0·554	280·9	27·9	+10·6	3	24	
			1300		521	0·536	280·4	26·8	+10·1	8	22	
(134)	(559)		(3499)		522	0·349	315·5	9·5	+19·6	53	290	
			542		522	0·317	319·0	7·1	+19·1	0	6	
			339		525	0·576	58·9	322·5	+21·0	4	8	
+14·2	16	98			525	0·621	56·9	319·8	+24·2	76	323	
+14·5	41	108	411 p		525	0·657	57·6	317·0	+24·9	14	124	306 c
+15·2	34	122			526	0·726	58·2	311·1	+26·4	34	179	
+11·4	0	15			526	0·755	58·5	308·4	+27·0	63	249	485 c
+13·3	0	15			527	0·692	115·9	314·9	−13·4	6	17	
+11·1	0	11			527	0·726	114·4	311·9	−13·4	0	11	
+19·3	114	402	442 f		528	0·773	113·8	307·8	−14·5	0	2	
			1043		528	0·801	112·4	304·9	−14·2	29	113	
(205)	(771)		(2777)		528	0·837	112·7	301·4	−15·6	20	59	
			477		528	0·855	111·7	299·2	−15·3	14	94	
+14·3	0	75	730 n p	July 25	528	0·880	112·2	296·5	−16·6	34	130	
+14·9	22	125			528	0·892	110·8	294·7	−15·7	28	78	602 e
+19·6	68	328				0·884	70·3			(465)	(2036)	147
+24·3	70	251	792 p	206·035	524	0·939	286·3	54·5	+17·2	56	176	(1919)
+26·2	0	130			524	0·934	286·8	49·9	+17·6	6	18	
			792		524	0·926	286·3	48·6	+17·2	21	47	381 c
(160)	(909)		(2791)		521	0·797	277·9	33·3	+ 9·7	17	56	
			315		521	0·778	278·4	31·6	+10·0	0	10	
+18·1	0	12			521	0·744	279·8	28·5	+10·9	1	2	155 e
+17·4	7	35			521	0·773	283·6	31·1	+14·0	20	100	528 n f
+17·3	2	11			522	0·522	299·2	9·2	+19·6	38	246	
+18·0	2	22			525	0·384	40·7	324·8	+22·2	0	3	
+14·6	20	100			525	0·402	50·5	321·2	+20·0	1	4	
+ 9·9	0	25			525	0·456	45·0	319·8	+23·9	98	352	
+10·6	18	156			525	0·494	47·1	317·0	+24·7	7	49	
+19·5	51	256			529	0·398	67·9	318·2	+13·7	5	24	
+21·7	0	31			526	0·604	51·2	308·7	+26·9	135	552	393 f
+13·0	5	28			527	0·515	127·7	315·7	−13·3	4	11	
+24·5	47	376			527	0·555	124·7	312·5	−13·5	1	7	
					528	0·655	119·3	304·4	−14·2	32	126	
					528	0·696	119·1	301·4	−15·5	22	128	

Areas of Spots and Faculæ are expressed in Millionths of the Sun's visible Hemisphere.   † Melbourne photo.

MEASURES of POSITIONS and AREAS of SPOTS and FACULÆ upon the Sun's DISK on PHOTOGRAPHS taken in the YEAR 1881—*continued.*

Mean Solar Time.	No. of Group and Letter for Spot.	Distance from Centre in terms of Sun's Radius.	Position Angle from Sun's Axis.	Longitude.	Latitude.	Area of UMBRA for each Spot (and for Day).	Area of WHOLE Spot for each (and for Day).	Area for each Group (and for Day).
1881.			°	°	°			
206°·035	528	0·730	118·6	298·6	−16·3	0	15	
	528	0·782	116·3	293·6	−16·5	13	136	680 e
		0·826	66·7					720
July 26						(477)	(2062)	(2857)
206·715	521	0·883	278·2	33·6	+ 9·8	0	14	
	521	0·863	277·0	31·2	+ 8·9	0	7	
	521	0·859	278·2	30·7	+ 9·9	0	6	
	521	0·859	282·8	30·7	+13·8	10	29	
	522	0·632	293·9	8·9	+19·3	41	115	921 nf
	525	0·367	30·3	319·7	+23·9	64	295	
	525	0·401	33·8	317·1	+24·8	0	9	
	525	0·456	35·9	313·9	+27·0	0	4	
	525	0·473	37·1	312·6	+27·4	0	4	
	525	0·459	40·0	312·2	+25·9	6	35	
	525	0·495	37·1	311·5	+28·4	0	4	
	526	0·509	43·3	308·4	+26·8	94	439	
	529	0·271	54·7	318·5	+14·4	7	26	
	527	0·446	137·6	313·6	−13·9	0	18	
	527	0·454	133·0	311·8	−11·7	0	19	
	528	0·549	126·0	304·2	−13·8	38	80	
	528	0·589	126·1	301·8	−15·4	29	70	
	528	0·666	123·9	300·1	−14·9	0	31	
	528	0·694	121·0	293·1	−16·6	39	120	
July 27*						(328)	(1334)	(921)
207·912		0·780	245·8					251
	521	0·977	275·4	33·8	+10·4	28	158	
	521	0·964	283·3	30·8	+14·3	0	15	
	521	0·950	279·6	27·8	+10·9	0	19	533 c
	522	0·803	290·5	8·4	+19·8	14	58	278 nf
	530	0·294	335·8	322·6	+ 9·9	0	8	
	525	0·321	349·3	319·3	+23·9	61	323	
	525	0·366	11·4	310·9	+25·6	0	6	
	529	0·137	340·6	318·2	+13·0	2	12	
	526	0·390	19·4	307·2	+27·1	129	511	
	527	0·339	172·6	313·0	−13·9	0	20	
	528	0·380	149·1	304·0	−13·4	14	81	
	528	0·428	146·8	301·6	−15·4	22	88	
	528	0·504	138·9	295·4	−16·9	15	83	
	531	0·374	54·8	296·9	+17·8	31	157	
	532	0·443	58·3	292·2	+18·6	18	106	
	533	0·433	73·2	246·0	+17·7	7	19	631 n
		0·850	56·3					232
July 28						(341)	(1664)	(1925)
208·898		0·857	248·7					541
	522	0·907	289·3	7·8	+19·9	6	15	710 nf
	525	0·409	320·1	319·1	+23·7	70	300	
	525	0·397	335·0	313·3	+26·6	0	13	
	534	0·478	210·6	317·3	−18·7	8	24	
	526	0·375	350·0	306·7	+27·3	126	490	
	528	0·354	187·7	305·3	−14·7	1	6	
	528	0·320	182·8	303·4	−11·8	8	46	
	528	0·365	181·6	303·1	−15·6	19	67	
	528	0·351	176·6	301·3	−14·7	12	42	
	528	0·377	169·9	298·6	−16·0	4	21	

Mean Solar Time.	No. of Group and Letter for Spot.	Distance from Centre in terms of Sun's Radius.	Position Angle from Sun's Axis.	Longitude.	Latitude.	Area of UMBRA for each Spot (and for Day).	Area of WHOLE Spot for each (and for Day).	Area for each Group (and for Day).
1881.			°	°	□			
208°·898	528	0·402	164·0	295·9	−16·9	7	50	
	531	0·224	19·4	298·1	+17·9	45	201	
	532	0·279	37·4	192·3	+18·4	44	209	
	533	0·833	72·2	246·3	+18·0	7	15	
	533	0·872	71·4	241·8	+19·0	0	16	507 c
		0·745	49·7					187
		0·950	112·8					190
July 29						(357)	(1515)	(2135)
209·953		0·956	290·3					616
	534	0·643	231·4	320·4	−18·7	6	37	
	534	0·620	229·0	318·0	−18·9	2	15	
	525	0·565	304·6	318·9	+23·7	42	244	
	526	0·498	319·7	309·7	+27·7	4	19	
	526	0·459	324·3	306·0	+27·3	93	457	
	528	0·452	215·5	304·3	−16·0	4	22	
	528	0·414	211·8	301·5	−14·9	27	123	
	528	0·402	198·7	296·2	−16·5	18	54	
	531	0·273	321·2	298·8	+18·0	60	224	
	528	0·244	325·7	296·7	+17·4	9	22	
	531	0·242	334·4	294·8	+18·3	19	68	
	531	0·209	340·3	292·7	+17·1	8	27	
	531	0·122	345·4	291·8	+18·1	25	136	
	533	0·686	69·4	246·1	+18·3	0	14	
	536	0·779	65·3	238·6	+22·8	0	16	
	536	0·802	61·5	236·9	+26·2	0	28	
	535	0·846	117·4	235·9	−19·4	3	31	139 c
	535	0·885	116·8	231·3	−20·4	0	32	
		0·795	65·1					299
July 30*						(320)	(1569)	(1054)
211·935	534	0·897	247·3	322·2	−17·3	53	145	235 f
	525	0·834	295·2	317·9	+24·2	76	265	1015 s
	526	0·716	303·7	304·5	+27·8	132	573	489 c
	528	0·734	242·4	304·6	−15·5	19	66	
	529	0·634	135·2	295·0	−16·2	0	46	
	531	0·630	294·0	299·8	+19·5	42	191	
	531	0·575	293·7	295·9	+18·3	7	26	
	532	0·521	297·5	291·5	+19·1	35	216	
	535	0·573	136·5	237·8	−19·1	5	21	
	535	0·591	135·0	236·2	−19·3	7	28	
	535	0·635	131·5	232·1	−19·7	0	18	
	536	0·536	47·5	236·2	+26·5	0	23	
	537	0·903	76·0	197·4	+15·2	0	21	918 nf
	538	0·988	77·0	180·4	+13·7	0	218	214 c
		0·925	114·8					314
Aug. 1						(376)	(1857)	(3185)
213·001		0·815	246·0					1067
		0·724	291·5					445
	534	0·974	251·1	322·3	−16·9	0	94	436 p
	525	0·935	293·3	317·9	+23·9	106	288	531 c
	526	0·841	299·2	304·0	+27·7	118	522	524 c
	528	0·876	248·3	305·7	−15·7	0	24	
	528	0·779	244·1	247·8	−15·8	3	27	
	531	0·798	290·1	300·7	+19·6	36	185	
	532	0·693	291·9	291·0	+19·4	12	117	

* Indian photo.   The Areas of Spots and Faculæ are expressed in Millionths of the Sun's visible Hemisphere.

MEASURES of POSITIONS and AREAS of SPOTS and FACULÆ upon the SUN's DISK on PHOTOGRAPHS taken in the YEAR 1881—*continued.*

Mean Solar Time.	No. of Group and Letter for Spot.	Distance from Centre in terms of Sun's Radius.	Position Angle from Sun's Axis.	HELIOGRAPHIC Longitude.	HELIOGRAPHIC Latitude.	SPOTS. Area of UMBRA for each Spot (and for Day).	SPOTS. Area of WHOLE Spot (and for Day).	FACULÆ. Area for each Group (and for Day).
1881.								
13ʰ·001	535	0·457	158·0	237·9	−19·0	10	26	
	536	0·370	29·6	236·7	+24·6	1	5	
	537	0·777	75·0	197·3	+15·5	0	20	972 n p
	538	0·923	78·2	180·4	+13·2	87	588	484 c
		0·865	120·8					670
Aug. 2						(373)	(1896)	(4929)
213·616		0·985	251·8					1021
		0·882	249·2					760
	525	0·975	293·0	318·5	+23·7	73	218	1056 c
	526	0·905	297·3	304·8	+27·3	88	394	287 c
	531	0·882	288·6	302·3	+19·3	48	207	276 c
	532	0·789	289·5	291·9	+19·1	21	80	
	539	0·833	128·5	193·3	−27·1	0	27	
	538	0·846	78·2	182·1	+13·3	61	407	
	540	0·876	115·0	183·6	−18·4	0	10	
	540	0·913	113·8	178·5	−18·8	0	20	
		0·874	124·7					637
		0·913	74·4					998
		0·937	114·7					580
Aug. 3*						(291)	(1363)	(5615)
214·951		0·973	251·3					796
	526	0·985	296·6	304·1	+27·2	95	449	227 e
	531	0·979	288·0	301·8	+18·8	31	153	
	532	0·926	287·7	290·8	+18·6	0	79	442 n f
	539	0·700	140·9	192·5	−27·3	1	7	173 f
	540	0·740	122·4	181·3	−18·7	21	54	266 c
	538	0·674	76·9	180·3	+13·3	57	398	510 n f
Aug. 4						(205)	(1140)	(2414)
216·061		0·968	292·5					203
		0·797	305·3					114
	541	0·384	325·2	221·7	+24·3	3	10	
	540	0·583	134·8	182·0	−18·6	14	68	
	538	0·464	74·2	180·7	+12·8	49	225	
	538	0·495	76·0	178·5	+12·3	17	102	647
		0·893	120·2					
Aug. 5						(83)	(403)	(964)
216·998		0·897	302·5					247
		0·807	291·7					182
	541	0·512	308·2	221·4	+24·1	2	16	
	540	0·476	152·3	182·0	−18·7	15	104	
	538	0·274	65·4	180·7	+12·6	38	210	
	538	0·293	70·2	179·0	+11·8	6	74	
	542	0·990	110·0	116·5	−18·6	0	41	591 s p
Aug. 6						(61)	(445)	(1020)
219·056		0·948	246·7					386
		0·910	293·8					448
	540	0·493	214·5	185·2	−17·9	15	85	
	540	0·460	207·1	181·8	−19·6	0	10	
	538	0·237	188·9	181·3	+10·6	1	9	
	538	0·237	298·4	180·5	+12·7	45	211	

Mean Solar Time.	No. of Group and Letter for Spot.	Distance from Centre in terms of Sun's Radius.	Position Angle from Sun's Axis.	HELIOGRAPHIC Longitude.	HELIOGRAPHIC Latitude.	SPOTS. Area of UMBRA for each Spot (and for Day).	SPOTS. Area of WHOLE Spot (and for Day).	FACULÆ. Area for each Group (and for Day).
1881.								
219·056	543	0·459	202·8	179·9	−18·6	29	158	
	542	0·845	117·4	115·8	−19·0	5	12	295 f
		0·885	76·3					483
Aug. 8						(93)	(485)	(1612)
220·068		0·797	291·2					450
		0·796	240·2					186
	540	0·632	230·9	185·7	−17·9	4	42	
	538	0·440	286·3	180·4	+12·9	40	197	
	543	0·567	223·5	178·9	−18·4	11	54	
	543	0·531	218·4	175·1	−18·5	2	6	
	543	0·515	214·2	172·6	−19·0	1	2	
	542	0·725	124·3	115·6	−19·1	6	13	330 f
		0·808						238
		0·880						214
Aug. 9						(64)	(314)	(1418)
221·094		0·885	290·2					893
	538	0·628	282·9	180·0	+13·1	51	179	
	543	0·707	235·2	178·9	−18·6	12	89	469 p
	544	0·718	133·1	106·4	−24·0	0	4	
		0·734	132·8					200
		0·963	123·5					304
Aug. 10						(63)	(272)	(1866)
221·889	538	0·755	281·7	179·9	+13·1	30	172	971 p
	543	0·796	242·2	178·1	−17·4	0	7	
	545	0·578	180·5	131·2	−28·6	0	5	
	546	0·587	176·2	128·2	−29·2	5	20	
	546	0·592	172·9	125·9	−29·3	21	37	
	542	0·498	149·5	115·3	−19·0	0	6	
		0·811	151·3					276
		0·929	124·5					194
		0·964	73·7					292
Aug. 11						(56)	(247)	(2582)
222·780		0·934	247·6					746
	538	0·868	281·0	179·7	+12·8	49	171	1148 p
	546	0·591	190·3	125·9	−28·8	10	51	
		0·912	71·1					635
Aug. 12*						(59)	(221)	(2519)
223·771		0·978	250·1					1042
		0·917	287·1					653
	538	0·957	281·1	179·8	+11·6	44	128	445
	540	0·638	205·5	134·0	−28·7	2	64	
		0·910	82·3					470
Aug. 13*						(46)	(192)	(2610)
224·805		0·965	291·3					678
		0·926	241·4					302
Aug. 14*								(980)
225·991		0·878	238·0					831
		0·909	69·7					189
		0·949	111·5					153
Aug. 15								(1295)

* Indian photo.   The Areas of Spots and Faculæ are expressed in Millionths of the Sun's visible Hemisphere.

MEASURES OF POSITIONS and AREAS of SPOTS and FACULÆ upon the SUN's DISK on PHOTOGRAPHS taken in the YEAR 1881—*continued.*

Mean Solar Time.	No. of Group, and Letter for Spot.	Distance from Centre in terms of Sun's Radius.	Position Angle from Sun's Axis.	HELIOGRAPHIC Longitude.	Latitude.	SPOTS Area of UMBRA for each Spot (and for Day).	Area of WHOLE for each Spot (and for Day).	FACULÆ Area for each Group (and for Day).
1881.								
227ᵈ·034		0·876	237·0					53z
		0·753	288·3					139
		0·894	111·8					255
Aug. 16								(941)
227·923		0·920	235·5					929
		0·805	287·7					351
	547	0·235	41·9	41·6	+16·8	9	26	
	547	0·275	50·5	38·2	+16·7	5	10	
		0·835	123·5					263
		0·878	109·7					182
		0·979	68·0					267
Aug. 17						(14)	(36)	(1992)
228·784	547	0·169	347·5	41·8	+16·3	0	33	
	547	0·170	6·2	38·5	+16·6	0	45	
Aug. 18*						(0)	(78)	
229·816		0·966	286·2					562
		0·724	270·3					130
		0·705	204·1					231
	547	0·282	310·8	38·9	+17·3	2	34	
		0·956	65·2					386
Aug. 19*						(2)	(34)	(1309)
231·040		0·915	231·8	38·9	+17·6	5	19	299
	547	0·507	293·2					
	548	0·925	108·3	305·2	−14·0	0	11	1081 s
	549	0·958	62·1	295·3	+28·7	51	194	929 s p
Aug. 20						(56)	(224)	(2309)
232·891		0·868	286·3					303
	549	0·756	60·6	298·1	+26·6	14	44	
	549	0·776	60·0	296·3	+27·5	3	32	
	549	0·795	59·0	294·6	+28·7	30	131	390 c
	550	0·948	77·7	273·0	+13·9	44	215	422 c
		0·772	18·7					482
Aug. 22						(91)	(422)	(1597)
233·902		0·902	284·7					464
	549	0·612	34·5	298·2	+26·7	2	22	
	549	0·661	53·9	294·7	+28·6	22	105	
	549	0·665	56·1	293·7	+27·3	7	53	
	550	0·847	78·4	273·6	+13·6	23	113	
	550	0·880	78·7	269·7	+13·3	11	21	
	551	0·899	73·3	267·3	+18·0	0	6	
	552	0·903	75·4	266·7	+16·2	9	53	
	552	0·909	77·1	265·9	+14·6	17	58	
	553	0·932	71·6	262·3	+19·7	73	277	
	554	0·935	71·1	257·3	+20·0	0	64	793 c
		0·900	123·5					204
		0·988	107·0					336
Aug. 23						(164)	(772)	(1797)

Mean Solar Time.	No. of Group, and Letter for Spot.	Distance from Centre in terms of Sun's Radius.	Position Angle from Sun's Axis.	HELIOGRAPHIC Longitude.	Latitude.	SPOTS Area of UMBRA for each Spot (and for Day).	Area of WHOLE for each Spot (and for Day).	FACULÆ Area for each Group (and for Day).
1881.								
234ᵈ·899		0·948	284·0					253
		0·820	242·3					323
		0·804	291·7					157
	549	0·469	42·0	298·2	+27·0	0	10	
	549	0·491	47·2	295·3	+25·9	10	42	
	549	0·537	44·4	194·1	+28·5	47	260	
	549	0·534	36·7	296·8	+31·8	4	15	
	550	0·703	77·8	274·0	+13·6	17	89	
	550	0·739	78·6	270·9	+13·2	13	65	
	552	0·783	75·2	267·0	+16·0	3	36	
	552	0·797	77·0	265·6	+14·7	22	104	
	553	0·825	70·2	263·2	+20·3	60	299	
	554	0·882	70·5	256·3	+20·5	6	31	
	554	0·905	70·7	253·3	+20·5	0	10	
		0·817	73·7					977
		0·952	109·7					635
Aug. 24						(182)	(961)	(2345)
235·771		0·903	244·7					300
	554a	0·271	187·6	309·4	−8·5	0	24	
	549	0·373	27·7	296·2	+26·2	11	53	
	549	0·381	32·0	294·4	+25·7	4	19	
	549	0·424	28·8	293·9	+28·6	32	206	
	549	0·440	33·2	291·5	+28·4	9	43	
	550	0·539	75·8	274·9	+13·6	10	58	
	550	0·554	77·4	273·8	+12·9	2	19	
	550	0·580	78·0	271·9	+12·7	6	26	
	550	0·611	76·6	269·7	+13·8	0	37	
	552	0·641	74·7	267·6	+15·2	2	42	
	552	0·666	75·7	265·6	+14·8	33	159	
	553	0·692	67·1	264·5	+20·8	45	292	
	554	0·756	66·3	259·0	+22·5	0	25	
	554	0·765	69·3	257·7	+20·4	0	22	
		0·899	114·0					698
Aug. 25*						(155)	(1025)	(998)
236·876		0·926	251·5					185
	555	0·287	303·0	307·1	+15·9	4	12	
	549	0·352	348·5	297·2	+27·2	18	46	
	549	0·325	335·0	294·5	+25·9	6	6	
	549	0·372	358·4	293·3	+28·9	25	194	
	556	0·374	40·1	277·5	+23·4	0	15	
	556	0·419	42·4	274·6	+24·7	9	12	
	550	0·331	69·7	274·2	+13·3	12	53	
	550	0·355	73·4	272·4	+12·5	8	34	
	550	0·376	72·3	271·2	+13·2	3	7	
	550	0·401	72·7	269·6	+13·5	0	5	
	550	0·417	72·1	268·7	+13·9	0	9	
	552	0·447	69·3	267·1	+15·5	3	28	
	552	0·480	72·5	264·6	+14·6	27	112	
	553	0·516	61·6	263·8	+20·4	89	328	
	553	0·553	62·6	261·1	+20·8	24	96	
	554	0·618	65·0	256·0	+20·9	18	57	
	557	0·995	108·4	211·3	−17·4	25	98	
		0·776	119·8					792
		0·875	66·0					688
Aug. 26						(265)	(1124)	(1663)

* Indian photo.   The Areas of Spots and Faculæ are expressed in Millionths of the Sun's visible Hemisphere.

MEASURES OF POSITIONS and AREAS of SPOTS and FACULÆ upon the SUN'S DISK on PHOTOGRAPHS taken in the YEAR 1881—*continued.*

Mean Solar Time.	No. of Group, and Letter for Spot.	Distance from Centre in terms of Sun's Radius.	Position Angle from Sun's Axis.	Heliographic Spots. Longitude.	Latitude.	Area of UMBRA for each Spot (and for Day).	Area of WHOLE Spot (and for Day).	Faculæ. Area for each Group (and for Day).
1881.								
237d·890		0·907	131·8					167
		0·752	292·3					48
	549	0·443	320·0	297·8	+26·5	12	29	
	549	0·426	331·1	292·8	+28·8	36	197	
	549	0·418	336·7	290·2	+29·5	0	26	
	556	0·272	3·3	278·3	+22·8	20	71	
	556	0·295	4·4	277·9	+24·2	10	26	
	556	0·299	9·1	276·3	+24·3	8	20	
	556	0·284	15·7	274·5	+22·9	7	13	
	556	0·326	13·5	274·5	+25·5	37	163	
	550	0·157	35·9	274·6	+13·5	9	40	
	550	0·138	45·6	273·5	+17·6	0	8	
	550	0·184	52·7	270·7	+13·5	15	71	
	550	0·221	62·1	267·8	+12·9	5	15	
	552	0·254	57·0	266·6	+14·9	14	111	
	552	0·292	53·5	265·1	+16·9	4	6	
	552	0·272	66·1	264·5	+13·2	5	36	
	553	0·283	61·0	264·5	+14·8	22	79	
	553	0·344	46·1	264·0	+20·7	81	350	
	553	0·380	49·4	261·3	+21·1	16	69	
	554	0·458	56·6	255·2	+21·1	9	43	
	557	0·942	111·2	112·7	−17·1	39	91	152 c
		0·796	64·3					126
		0·921	71·3					469
Aug. 27						(349)	(1464)	(962)
238·899		0·839	294·8					558
		0·700	237·2					429
	549	0·584	304·9	297·9	+25·6	0	3	
	549	0·545	314·0	292·4	+28·6	33	176	
	556	0·538	322·5	278·7	+22·5	9	67	
	556	0·326	331·2	275·8	+23·6	0	13	
	556	0·337	339·7	273·9	+26·6	21	42	
	556	0·312	336·9	273·6	+23·7	11	20	
	556	0·328	340·1	273·0	+25·0	13	73	
	550	0·177	305·1	274·4	+12·9	2	20	
	550	0·140	318·4	271·4	+13·1	24	141	
	552	0·152	334·9	269·7	+15·0	4	14	
	552	0·182	336·9	270·2	+16·8	10	16	
	552	0·107	344·6	267·6	+13·1	9	63	
	552	0·118	1·0	265·8	+13·9	49	313	
	553	0·233	6·5	264·3	+20·5	81	300	
	553	0·252	15·9	261·7	+21·1	15	70	
	553	0·305	36·1	251·9	+21·2	0	6	
	557	0·846	115·4	213·1	−17·0	20	95	375 c
								522
Aug. 28		0·880	71·8			(301)	(1442)	(1884)
239·812		0·923	293·3					546
		0·907	296·7					574
	549	0·666	305·6	291·8	+28·4	26	152	
	556	0·478	303·8	279·1	+21·9	17	79	
	556	0·462	306·5	277·5	+22·5	0	4	
	556	0·470	309·0	277·3	+23·8	0	6	
	556	0·449	319·8	272·8	+26·8	5	27	
	556	0·424	316·5	272·6	+24·7	16	72	
	550	0·326	297·9	271·2	+15·7	17	60	

Mean Solar Time.	No. of Group, and Letter for Spot.	Distance from Centre in terms of Sun's Radius.	Position Angle from Sun's Axis.	Heliographic Spots. Longitude.	Latitude.	Area of UMBRA for each Spot (and for Day).	Area of WHOLE Spot (and for Day).	Faculæ. Area for each Group (and for Day).
1881.								
239d·812	550	0·309	288·5	271·3	+12·5	15	70	
	552	0·298	299·8	269·4	+15·4	7	24	
	552	0·368	303·4	267·3	+15·5	19	72	
	552	0·255	295·8	267·5	+13·4	59	201	
	552	0·218	308·0	264·1	+14·8	18	82	
	552	0·171	332·8	258·5	+15·9	0	13	
	553	0·287	313·2	264·4	+20·3	36	262	
	553	0·275	330·3	262·3	+20·9	7	47	
	553	0·229	355·3	255·0	+20·4	0	58	
	559	0·428	186·4	256·8	−17·9	0	14	
	559	0·428	179·9	253·8	−18·1	2	18	
	557	0·738	122·0	213·0	−17·6	2	48	
	561	0·989	117·3	177·2	−25·5	0	139	1442 c
		0·903	79·4					412
Aug. 29*						(246)	(1448)	(2974)
241·025		0·602	246·0					476
	549	0·818	299·9	291·4	+28·4	25	157	
	555	0·686	294·3	280·0	+21·7	24	146	
	556	0·671	299·8	277·7	+25·0	7	28	
	556	0·654	295·5	277·2	+21·9	0	6	
	556	0·651	303·8	273·6	+26·4	12	74	
	550	0·551	282·0	271·3	+12·6	5	46	
	551	0·489	286·5	266·7	+14·3	78	593	
	553	0·472	300·3	263·5	+20·3	69	388	
	558	0·395	294·6	259·7	+16·1	23	116	
	559	0·502	211·9	254·5	−17·9	4	16	
	557	0·579	135·6	211·8	−17·9	1	30	
	560	0·870	78·5	176·9	+13·5	9	50	315 p
	561	0·928	120·9	176·3	−25·2	103	431	1310 e
		0·900	68·2					269
Aug. 30						(366)	(2091)	(3503)
242·907	549	0·970	297·7	290·6	+28·6	39	144	564 c
	556	0·002	291·0	278·0	+22·9	10	53	554 c
	552	0·802	282·2	266·7	+14·8	65	510	1090 c
	553	0·771	290·2	263·2	+20·1	88	384	
	559	0·741	236·7	253·4	−18·9	17	73	320 p
	561	0·765	136·9	183·1	−24·8	106	507	
	561	0·768	133·0	174·5	−23·9	30	168	369 c
	560	0·594	77·3	176·6	+13·4	5	24	
	562	0·783	113·2	168·9	−20·2	0	22	197 f
		0·730	68·5					91
		0·942	118·7					167
Sept. 1						(360)	(1875)	(3352)
244·549		0·898	247·1					739
	552	0·963	284·6	266·7	+15·9	0	258	1755 n f
	553	0·944	284·6	263·2	+20·9	69	261	
	561	0·591	165·6	181·0	−27·0	24	97	
	561	0·540	163·8	181·9	−23·5	51	206	
	561	0·505	162·8	180·7	−25·4	14	83	
	561	0·603	157·2	176·2	−26·7	0	84	
	561	0·619	150·3	171·3	−25·6	0	27	
Sept. 3†						(158)	(1016)	(2494)

* Indian photo.　　　The Areas of Spots and Faculæ are expressed in Millionths of the Sun's visible Hemisphere.　　　† Melbourne photo.

Measures of Positions and Areas of Spots and Faculæ upon the Sun's Disk on Photographs taken in the Year 1881—*continued.*

Mean Solar Time	No. of Group, and Letter for Spot	Distance from Centre in terms of Sun's Radius	Position Angle from Sun's Axis	Heliographic Longitude	Latitude	Area of UMBRA for each Spot (and for Day)	Area of WHOLE Spot (and for Day)	Area for each Group (and for Day)
1881.								
246ᵈ·155		0·966	289·5					300
	561	0·557	198·5	181·2	−24·7	51	307	
	561	0·587	196·2	180·7	−27·0	31	148	
	561	0·554	193·1	178·1	−25·3	0	25	
	561	0·546	185·5	173·4	−25·5	14	111	
	563	0·309	189·9	173·2	−10·4	0	17	
		0·888	127·0					155
		0·961	75·5					302
Sept. 4						(96)	(608)	(757)
246·789	561	0·606	209·8	181·0	−24·9	49	275	
	561	0·626	207·7	180·7	−26·8	42	152	
	561	0·595	203·8	177·1	−25·9	10	56	
	561	0·595	200·8	175·3	−26·7	2	28	
	561	0·563	198·9	173·3	−25·0	20	118	
	563	0·385	219·6	176·1	−10·2	21	51	
	563	0·351	208·4	171·4	−10·8	12	35	
		0·980	166·6					510
		0·953	76·3					2006
Sept. 5*						(156)	(715)	(2526)
247·936		0·933	248·7					128
	564	0·950	289·8	219·6	+21·1	15	45	
	563	0·661	283·4	188·0	+14·3	11	25	
	561	0·723	224·4	180·4	−25·0	58	283	
	561	0·736	221·9	180·0	−27·1	12	108	
	561	0·692	218·0	173·8	−25·5	16	126	759 c
	563	0·574	241·2	177·2	− 9·8	11	57	
	563	0·513	232·1	171·1	−11·7	5	19	
	566	0·717	78·3	99·7	+13·5	13	40	382 f
		0·893	134·3					316
Sept. 6						(141)	(703)	(2355)
249·067		0·923	291·7					174
	565	0·851	282·5	190·5	+14·4	0	9	162 f
	561	0·833	336·6	180·4	−22·5	0	9	
	561	0·840	233·9	180·0	−24·9	61	289	
	561	0·846	231·8	176·6	−26·7	31	97	
	561	0·794	230·2	173·9	−25·2	0	87	884 c
	563	0·755	250·5	177·7	− 9·5	10	25	
	566	0·513	76·7	101·0	+13·0	49	141	
	566	0·534	76·0	99·5	+13·6	7	42	
	566	0·561	73·5	97·9	+15·2	15	61	
	567	0·713	74·4	86·2	+16·2	2	18	184 c
Sept. 7						(175)	(778)	(1404)
249·738		0·357	251·2					552
		0·957	243·1					938
		0·919	182·8					530
		0·919	290·0					367
	561	0·900	237·6	179·6	−25·0	63	222	1569 c
	561	0·853	234·0	172·5	−25·5	15	62	
	566	0·370	73·2	101·6	+12·9	21	123	

Mean Solar Time	No. of Group, and Letter for Spot	Distance from Centre in terms of Sun's Radius	Position Angle from Sun's Axis	Heliographic Longitude	Latitude	Area of UMBRA for each Spot (and for Day)	Area of WHOLE Spot (and for Day)	Area for each Group (and for Day)
1881.								
49ᵈ·738	566	0·378	70·0	101·4	+14·1	0	18	
	566	0·393	74·2	100·1	+12·8	19	51	
	566	0·449	71·4	96·8	+14·7	27	78	
	567	0·601	73·6	86·2	+15·6	0	14	
Sept. 8*						(145)	(568)	(3956)
250·769		0·910	253·7					427
		0·907	292·4					593
	561	0·969	241·9	179·5	−24·9	65	194	3078 c
	566	0·149	46·5	102·9	+13·0	28	255	
	566	0·209	59·1	98·6	+13·3	0	63	
	566	0·240	63·6	96·5	+13·1	10	27	
	566	0·262	57·4	96·0	+15·1	32	202	
		0·878	80·4					236
Sept. 9*						(135)	(741)	(4334)
251·688		0·980	240·3					1760
		0·971	256·6					468
		0·961	292·2					754
	566	0·151	318·5	102·9	+13·7	66	271	
	566	0·122	327·1	100·9	+13·0	2	33	
	566	0·106	336·7	99·5	+12·8	5	18	
	566	0·111	347·5	98·5	+14·0	0	16	
	566	0·102	10·6	95·9	+13·0	16	67	
	566	0·149	7·7	95·8	+15·7	35	148	
	566	0·135	12·6	95·3	+14·7	0	48	
Sept.10*						(124)	(601)	(2982)
252·666		0·931	238·5					484
	566	0·350	289·2	104·0	+13·4	57	356	
	566	0·308	290·9	101·3	+13·2	2	111	
	566	0·285	294·7	99·6	+13·8	20	85	
	566	0·260	296·9	97·9	+13·7	19	160	
	566	0·266	309·8	96·5	+16·8	19	148	
	569	0·835	73·5	27·2	+17·7	12	32	
	570	0·865	71·2	23·9	+19·8	10	48	
Sept.11*						(149)	(940)	(484)
253·787	566	0·566	282·8	103·8	+13·2	65	304	
	566	0·566	285·9	103·5	+14·9	0	55	
	566	0·534	283·2	101·5	+13·1	0	38	
	566	0·525	285·0	100·8	+14·0	19	67	
	566	0·504	285·1	99·3	+13·8	0	19	
	566	0·487	282·6	98·4	+12·4	0	13	
	566	0·496	292·2	98·0	+17·1	28	103	
	566	0·476	285·6	97·4	+13·7	0	61	
	566	0·464	286·3	96·6	+13·9	9	35	
	566	0·454	285·3	96·0	+13·3	3	16	
	566	0·454	287·9	95·8	+14·5	0	7	
	568	0·383	244·4	89·5	− 2·8	18	53	
	568	0·333	238·7	85·9	− 3·1	26	95	
	569	0·669	72·6	27·7	+17·0	4	24	
	570	0·733	70·9	22·5	+18·9	12	36	

* Indian photo.   The Areas of Spots and Faculæ are expressed in Millionths of the Sun's visible Hemisphere.

MEASURES of POSITIONS and AREAS of SPOTS and FACULÆ upon the SUN's DISK on PHOTOGRAPHS taken in the Year 1881—*continued*.

Mean Solar Time.	No. of Group, and Letter for Spot.	Distance from Centre in terms of Sun's Radius.	Position Angle from Sun's Axis.	HELIOGRAPHIC		SPOTS		FACULÆ
				Longitude.	Latitude.	Area of UMBRA for each Spot (and for Day).	Area of WHOLE for each Spot (and for Day).	Area for each Group (and for Day).
1881.		°	°	°	°			
253ᵈ·787	571	0·970	75·7	352·3	+15·6	0	131	
	571	0·976	74·5	350·6	+16·7	0	139	
	572	0·976	69·4	350·4	+21·6	0	154	1346 c
	572	0·985	71·4	347·6	+19·5	0	156	
Sept. 12*						(183)	(1506)	(2346)
255·083	566	0·773	280·9	103·2	+13·0	126	553	
	566	0·710	286·5	98·2	+16·8	24	129	
	566	0·703	282·5	97·0	+13·9	20	95	544 c
	568	0·657	256·3	91·8	−3·4	22	103	
	568	0·570	253·4	85·3	−3·3	41	113	
	569	0·444	65·2	27·4	+17·2	5	13	
	570	0·536	64·3	21·5	+19·6	10	30	
	571	0·864	74·5	351·9	+17·0	76	224	
	572	0·887	68·2	349·3	+22·6	86	238	
	572	0·906	69·6	346·6	+21·5	69	235	
	573	0·932	75·7	344·2	+17·8	0	37	548 c
		0·806	127·8					
Sept. 13						(479)	(1770)	(1352)
255·943		0·884	230·2					283
	566	0·882	280·0	103·3	+12·2	83	494	
	566	0·839	284·5	98·4	+16·0	28	115	493 c
	566	0·818	281·5	97·2	+13·5	9	46	
	568	0·798	259·6	92·6	−3·9	24	109	
	568	0·710	258·2	85·6	−3·4	25	107	255 c
	569	0·289	50·5	27·4	+17·6	0	2	
	570	0·381	53·0	21·9	+19·7	11	29	
	571	0·753	73·1	352·0	+17·4	52	225	
	572	0·795	66·9	348·7	+22·7	69	277	
	572	0·816	68·0	346·4	+22·1	25	230	
	573	0·836	73·1	343·8	+18·0	13	27	987 c
Sept. 14						(339)	(1661)	(2018)
257·059	566	0·972	280·6	103·4	+12·0	97	451	
	566	0·948	284·5	98·6	+16·0	4	17	} 463 nf
	566	0·945	281·5	97·9	+13·2	0	11	
	568	0·924	263·0	92·7	−3·7	34	125	
	568	0·871	261·9	85·7	−3·4	26	63	267 c
	570	0·226	17·3	12·1	+19·6	4	20	
	571	0·571	69·5	352·2	+17·5	56	239	
	572	0·650	63·4	347·3	+22·6	90	538	
	573	0·675	71·7	344·1	+17·6	10	19	
		0·873	66·0					547
		0·915	110·2					177
Sept. 15						(321)	(1483)	(2039)
258·102		0·961	284·7					445
		0·651	278·5					125
	568	0·990	265·3	93·4	−3·6	0	87	333 f
	571	0·382	60·1	352·1	+17·7	57	134	
	572	0·482	53·0	347·7	+23·3	53	294	
	572	0·492	57·4	346·0	+21·7	23	152	
	573	0·489	67·4	344·3	+17·1	4	13	
		0·878	114·3					250
		0·902	62·3					791
Sept. 16						(137)	(780)	(1944)

Mean Solar Time.	No. of Group, and Letter for Spot.	Distance from Centre in terms of Sun's Radius.	Position Angle from Sun's Axis.	HELIOGRAPHIC		SPOTS		FACULÆ
				Longitude.	Latitude.	Area of UMBRA for each Spot (and for Day).	Area of WHOLE for each Spot (and for Day).	Area for each Group (and for Day).
1881.		°	°	°	°			
258ᵈ·952	571	0·781	277·0					111
	571	0·244	39·2	351·9	+17·9	49	248	
	572	0·365	38·7	346·8	+23·4	75	451	
	572	0·353	49·2	344·7	+20·1	3	9	
	573	0·334	57·3	344·1	+17·2	0	5	
		0·892	113·3					223
		0·928	61·5					487
Sept. 17						(127)	(713)	(821)
259·738	573a	0·807	278·8	55·1	+11·0	18	69	649 c
	571	0·183	353·9	352·1	+17·6	31	106	
	571	0·175	358·4	351·2	+17·2	15	46	
	572	0·291	11·6	347·3	+23·6	34	235	
	572	0·239	15·8	346·9	+20·3	8	46	
	572	0·259	17·7	346·1	+21·4	8	33	
	572	0·279	18·8	345·3	+22·4	8	50	
	572	0·305	18·7	344·8	+23·8	5	64	
	572	0·252	26·4	344·1	+20·1	3	15	
		0·946	60·0					1230
		0·945	73·8					1537
Sept. 18*						(130)	(664)	(3416)
260·722		0·931	281·8					262
	573a	0·972	279·3	55·0	+10·7	27	134	538 c
	571	0·304	308·1	352·3	+17·6	30	135	
	571	0·287	309·4	352·2	+17·3	15	47	
	571	0·288	315·8	350·0	+18·8	0	13	
	572	0·321	332·5	347·1	+23·5	39	255	
	572	0·280	327·3	347·1	+20·6	2	30	
	572	0·292	336·3	345·1	+22·5	12	118	
	574	0·955	76·3	264·1	+15·1	38	107	
	575	0·979	69·6	257·8	+21·4	39	217	
		0·860	72·1					383
		0·866	74·0					495
		0·894	58·0					676
		0·933	78·9					1012
Sept. 19*						(202)	(1056)	(3366)
261·961		0·878	288·0					161
	571	0·521	292·9	351·6	+17·8	43	167	
	572	0·480	307·7	345·8	+23·5	61	269	
	574	0·839	75·6	264·0	+15·0	19	92	315 p
	575	0·891	69·5	257·9	+21·5	51	164	273 f
		0·761	57·8					331
		0·892	116·5					1347
Sept. 20						(174)	(792)	(1347)
262ᵈ·784	571	0·670	289·1	352·1	+18·0	30	160	
	571	0·648	288·4	350·4	+17·2	0	13	
	572	0·613	299·7	345·8	+23·3	36	231	
	574	0·711	76·5	264·3	+14·6	12	40	
	575	0·798	68·9	257·8	+21·0	48	228	
		0·883	69·7					748
		0·925	113·0					1176
Sept. 21*						(126)	(672)	(1924)

* Indian photo. The Areas of Spots and Faculæ are expressed in Millionths of the Sun's visible Hemisphere.

MEASURES of POSITIONS and AREAS of SPOTS and FACULÆ upon the Sun's Disk on PHOTOGRAPHS taken in the Year 1881—continued.

Mean Solar Time	No. of Group, and Letter for Spot	Distance from Centre in terms of Sun's Radius	Position Angle from Sun's Axis	HELIOGRAPHIC Longitude	HELIOGRAPHIC Latitude	SPOTS Area of UMBRA for each Spot (and for Day)	SPOTS Area of WHOLE Spot (and for Day)	FACULÆ Area for each Group (and for Day)
1881. 263ᵈ·787	571	0·810	287·1	351·6	+18·0	27	95	218 c
	572	0·756	295·7	345·2	+23·8	44	196	
	572	0·731	293·1	343·4	+21·5	0	13	352 c
	575a	0·320	328·0	307·8	+22·6	0	26	
	574	0·557	74·9	263·8	+14·1	10	35	
	575	0·654	66·2	257·7	+20·7	36	201	
Sept. 22*						(117)	(566)	(570)
264·787	571	0·921	286·6	352·0	+18·0	21	95	577 c
	572	0·872	293·5	344·8	+23·8	49	196	587 c
	574	0·372	68·1	263·3	+14·5	9	30	
	575	0·486	59·0	257·7	+20·8	35	239	(1164)
Sept. 23*						(114)	(560)	
265·923		0·947	234·0					122
		0·826	295·3					174
	571	0·991	286·8	353·0	+17·5	30	91	
	572	0·966	292·4	345·6	+23·3	32	205	525 c
	574	0·164	34·6	263·7	+14·6	8	24	
	575	0·303	36·4	258·1	+20·9	50	242	
		0·858	68·0					182
		0·964	74·7					202
Sept. 24						(120)	(562)	(1205)
266·952		0·989	236·7					303
		0·913	295·0					250
		0·853	241·7					211
	574	0·186	316·0	263·2	+14·5	6	14	
	575	0·243	352·1	257·6	+20·7	35	259	
	576	0·993	7·6	171·1	+15·1	72	348	340 p
		0·935	113·3					269
Sept. 25						(113)	(621)	(1373)
267·988		0·913	292·8					423
		0·888	240·7					321
	574	0·379	291·6	263·1	+14·4	10	16	
	575	0·350	313·6	257·5	+20·5	33	218	
	577	0·958	118·9	174·2	-25·2	0	9	632 c
	576	0·942	76·7	170·7	+14·8	65	312	578 nf
		0·786	74·3					115
Sept. 26						(108)	(555)	(2069)
269·057		0·918	241·0					194
		0·839	302·0					474
		0·772	287·8					383
	574	0·575	285·7	262·5	+14·5	0	5	
	575	0·524	298·6	257·0	+20·4	42	236	
	578	0·268	175·5	226·5	- 8·7	5	13	
	577	0·877	123·9	174·3	-25·3	0	16	743 f
	576	0·841	77·0	170·2	+14·6	49	264	699 nf
Sept. 27						(96)	(534)	(2493)

Mean Solar Time	No. of Group, and Letter for Spot	Distance from Centre in terms of Sun's Radius	Position Angle from Sun's Axis	HELIOGRAPHIC Longitude	HELIOGRAPHIC Latitude	SPOTS Area of UMBRA for each Spot (and for Day)	SPOTS Area of WHOLE Spot (and for Day)	FACULÆ Area for each Group (and for Day)
1881. 269ᵈ·951		0·869	293·8					836
		0·803	238·8					133
	574	0·729	283·8	262·7	+14·6	0	5	
	575	0·674	293·2	256·9	+20·5	54	255	
	578	0·337	221·2	228·7	- 8·1	4	7	
	577	0·794	128·9	173·0	-25·0	8	15	595 n
	576	0·717	75·9	170·0	+14·8	61	289	494 f
Sept. 28						(127)	(571)	(2058)
270·999		0·923	292·2					869
		0·895	243·8					228
	575	0·818	290·7	256·5	+20·7	58	264	255 p
	578	0·527	243·0	230·1	- 7·9	1	9	
	579	0·415	196·8	209·1	-16·7	6	27	
	580	0·196	37·1	195·0	+15·6	0	15	
	577	0·682	139·0	172·5	-15·1	7	25	567 sf
	576	0·538	72·3	170·0	+15·1	62	289	
		0·943	72·7					583
Sept. 29						(134)	(629)	(2502)
272·044	575	0·928	243·3					250
	579	0·926	289·1	256·8	+20·2	65	223	457 nf
	580	0·53a	225·1	211·7	-16·2	3	11	
	577	0·173	336·4	192·5	+15·7	9	14	
	576	0·58a	154·4	172·3	-25·0	3	8	
		0·894	64·5	170·1	+14·6	67	227	464
		0·923						113
Sept. 30						(147)	(483)	(1284)
272·896		0·935	246·2					82
		0·808	253·7					92
	575	0·975	289·6	255·6	+20·6	37	213	304 nf
	579	0·669	237·0	212·7	-16·0	0	2	
	580	0·337	299·8	194·8	+15·9	7	20	
	577	0·53a	171·5	172·2	-25·0	3	8	
	576	0·185	41·7	169·9	+14·5	36	223	
	581	0·826	121·1	128·1	-20·9	7	19	122 c
	582	0·963	79·3	102·0	+12·1	9	34	612 p
		0·796	72·3					268
		0·975	113·3					194
Oct. 1						(99)	(519)	(1674)
274·079		0·933	259·2					167
		0·838	240·2					152
	583	0·853	288·8	230·2	+19·4	15	57	267 c
	580	0·575	288·3	159·9	+15·8	5	24	
	576	0·205	313·2	170·4	+14·5	38	211	
	581	0·677	129·5	127·9	-20·0	13	43	
	582	0·856	80·0	102·3	+11·9	4	16	492 c
		0·907	116·8					123
		0·925	96·2					98
Oct. 2						(75)	(351)	(1299)

* Indian photo.   The Areas of Spots and Faculæ are expressed in Millionths of the Sun's visible Hemisphere.

MEASURES of POSITIONS and AREAS of SPOTS and FACULÆ upon the SUN'S DISK on PHOTOGRAPHS taken in the YEAR 1881—*continued.*

Mean Solar Time	No. of Group, and Letter for Spot	Distance from Centre in terms of Sun's Radius	Position Angle from Sun's Axis	HELIOGRAPHIC Longitude	HELIOGRAPHIC Latitude	Area of UMBRA for each Spot (and for Day)	Area of WHOLE Spot (and for Day)	Area for each Group (and for Day)
1881.								
174*·920		0·922	307·2					193
	583	0·926	183·6	118·9	+19·6	11	53	216 c
	579	0·911	249·4	212·9	−15·7	9	20	190 s
	580	0·703	285·9	194·9	+15·8	13	46	167 c
	576	0·360	293·9	170·7	+14·5	28	158	
	581	0·573	141·5	128·1	−20·5	9	47	
	582	0·740	79·6	102·5	+12·1	8	11	
	584	0·807	77·1	96·4	+14·2	0	14	508 c
		0·852	98·2					87
		0·868	121·0					146
Oct. 3						(78)	(349)	(1507)·
275·962								326
	583	0·969	250·5					
	580	0·979	289·9	216·3	+20·8	40	141	252 s p
	576	0·560	284·1	195·8	+15·4	8	56	361 c
	581	0·473	286·2	170·2	+14·4	24	141	
	582	0·560	161·9	127·7	−20·3	25	93	
	584	0·642	78·2	102·7	+11·9	8	13	
		0·983	75·5	97·0	+14·2	0	5	
			79·7					222
Oct. 4						(103)	(449)	(1161)
276·786	580	0·952	283·5	198·8	+14·8	0	88	
	580	0·921	283·3	193·5	+14·7	8	62	1277 c
	576	0·701	283·7	170·2	+14·1	20	85	
	581	0·446	191·2	131·1	−19·5	20	108	
	581	0·453	182·7	127·1	−20·4	16	60	
	581	0·485	176·7	124·1	−22·4	0	35	
	582	0·397	74·4	102·9	+12·1	4	18	590
		0·946	79·0					(1867)
Oct. 5*						(68)	(456)	
277·791		0·981	184·6					771
		0·903	235·1					1391
	576	0·844	283·0	170·5	+14·3	27	113	
	585	0·803	189·0	165·8	+19·0	5	54	
	581	0·530	216·0	131·8	−19·4	23	180	
	581	0·492	204·5	125·1	−20·3	17	65	
	581	0·524	200·1	123·9	23·2	0	15	1240 c
	586	0·959	76·7	38·2	+14·6	78	267	(3402)
Oct. 6*						(155)	(694)	
278·899		0·934	237·0					406
	576	0·950	283·0	170·5	+14·3	29	79	
	585	0·937	189·5	168·1	+20·4	0	50	556 n f
	581	0·669	232·3	131·8	−18·9	31	165	
	581	0·618	223·1	125·6	−20·3	20	109	143 c
	586	0·857	76·9	38·6	+14·4	87	343	
	586	0·883	79·3	35·5	+13·3	12	204	573 n f
		0·931	105·7					200
Oct. 7						(179)	(950)	(1878)

Mean Solar Time	No. of Group, and Letter for Spot	Distance from Centre in terms of Sun's Radius	Position Angle from Sun's Axis	HELIOGRAPHIC Longitude	HELIOGRAPHIC Latitude	Area of UMBRA for each Spot (and for Day)	Area of WHOLE Spot (and for Day)	Area for each Group (and for Day)
1881.								
279d·796		0·974	289·5					1445
	581	0·789	239·8	132·0	−19·0	28	155	637
	581	0·746	233·7	126·1	−21·5	10	90	
	581	0·705	228·3	120·7	−22·8	4	20	
	586	0·733	76·4	38·9	+14·2	47	280	
	586	0·737	79·2	38·4	+12·1	0	16	
	586	0·763	77·1	36·0	+13·8	5	57	
	586	0·802	78·1	31·5	+13·3	11	48	1145 s f
Oct. 8*						(105)	(666)	(3227)
280·800	581	0·903	245·0	132·9	−19·5	18	103	
	581	0·870	242·3	127·9	−20·4	0	122	
	581	0·851	241·6	125·5	−20·2	9	39	926 c
	581	0·856	239·4	125·3	−22·2	0	28	
	586a	0·726	222·2	105·0	−27·4	0	67	
	586	0·564	73·9	39·0	+14·1	38	269	
	586	0·594	75·6	36·7	+13·4	0	36	
	586	0·646	77·2	32·7	+12·9	4	21	474
		0·799	72·7					(1400)
Oct. 9*						(69)	(655)	
281·793	581	0·974	248·3	133·3	−19·5	0	138	
	581	0·952	246·2	127·8	−20·4	0	128	907 c
	581	0·936	245·4	124·7	−20·4	0	67	
	586	0·380	66·6	38·7	+14·4	43	295	
	586	0·432	73·6	34·7	+12·6	0	10	
	586	0·470	74·0	32·2	+12·8	0	18	
	587	0·502	116·2	35·4	−11·7	15	61	
	587	0·553	127·7	31·4	−11·0	4	57	
	588	0·973	66·4	341·9	+24·3	0	142	745 c
Oct. 10*						(62)	(886)	(1652)
282·784		0·979	246·4					840
		0·877	282·1					559
	586	0·193	42·8	39·0	+14·1	41	219	
	586	0·214	52·0	36·8	+13·5	0	29	
	587	0·342	150·4	36·8	−11·3	17	110	
	587	0·384	140·9	32·4	−11·4	0	48	541 s f
	587	0·408	141·7	31·7	−12·8	0	17	
	587	0·413	138·7	30·6	−11·3	17	70	
	588	0·915	66·0	340·4	+24·3	4	109	(1940)
Oct. 11*						(79)	(602)	
283·796		0·956	238·1					1367
		0·934	282·6					1562
	586	0·169	332·1	39·1	+13·7	48	242	
	586	0·192	340·9	37·0	+16·3	3	16	
	586	0·140	334·3	36·9	+13·1	0	29	
	587	0·309	193·4	37·5	−11·6	13	122	
	587	0·296	188·1	35·7	−11·1	3	15	
	587	0·308	176·4	32·2	−11·9	0	33	
	587	0·321	171·8	30·6	−11·6	17	72	
	588	0·814	63·9	340·1	+24·5	14	63	1124
		0·929	66·2					(4053)
Oct. 12						(108)	(592)	

* Indian photo.   The Areas of Spots and Faculæ are expressed in Millionths of the Sun's visible Hemisphere.

MEASURES of POSITIONS and AREAS of SPOTS and FACULÆ upon the SUN'S DISK on PHOTOGRAPHS take

Mean Solar Time.	No. of Group, and Letter for Spot.	Distance from Centre in terms of Sun's Radius.	Position Angle from Sun's Axis.	Heliographic Longitude.	Heliographic Latitude.	SPOTS. Area of UMBRA for each Spot (and for Day).	SPOTS. Area of WHOLE Spot (and for Day).	FACULÆ. Area for each Group (and for Day).
1881. 284ᵈ·780		0·952	288·2					1536
	586	0·348	294·5	39·2	+13·8	34	259	
	587	0·422	226·6	38·4	−11·2	9	35	
	587	0·408	224·6	37·2	−11·2	0	21	
	587	0·364	215·8	32·8	−11·3	0	24	
	587	0·354	209·9	30·6	−12·0	15	68	
	588	0·695	59·6	339·1	+25·0	0	24	
		0·847	66·1					670
Oct. 13*						(58)	(431)	(2206)
285·898	586	0·566	285·9	39·5	+13·7	56	244	
	587	0·617	243·1	39·6	−11·4	4	9	
	587	0·518	235·4	31·5	−11·8	10	19	
	588	0·533	51·3	338·5	+24·6	7	19	
	589	0·966	74·8	289·8	+16·1	30	314	372 n p
		0·917	113·3					211
Oct. 14						(107)	(605)	(583)
286·898	586	0·737	283·7	39·7	+13·9	45	227	
	586	0·711	287·6	37·2	+16·4	10	38	252 c
	589	0·881	75·0	290·4	+15·9	47	288	619 f
		0·878	59·2					133
Oct. 15						(102)	(553)	(1004)
287·787								512
	586	0·857	282·9	39·9	+14·0	43	184	998 c
	589	0·773	74·2	290·3	+15·7	38	170	
	589	0·840	74·6	283·3	+16·0	0	23	1152 n f
	591	0·976	69·9	262·1	+20·9	0	143	
	590	0·980	78·0	261·4	+12·9	0	196	1568 c
Oct. 16*						(81)	(716)	(4230)
288·923		0·948	255·2					297
	586	0·962	282·3	40·4	+13·3	53	308	393 n f
	589	0·590	69·9	290·6	+16·2	52	199	
	589	0·631	71·7	287·7	+15·8	5	157	
	589	0·682	72·8	283·3	+15·7	9	25	237 f
	590	0·88g	77·7	262·7	+13·5	32	220	291 n f
	591	0·893	69·2	262·5	+21·0	30	110	510 f
	592	0·931	113·0	260·9	−19·0	57	190	
	592	0·955	112·9	256·6	−19·9	0	8	406 c
		0·922	56·2					117
Oct. 17						(238)	(1217)	(2251)
289·896		0·916	290·8					170
	589	0·437	.63·9	288·9	+16·0	78	335	
	590	0·769	75·8	262·8	+14·4	46	223	
	591	0·783	67·6	262·3	+20·8	27	126	424 f
	592	0·831	117·8	261·9	−19·4	37	178	
	592	0·874	116·6	256·8	−20·0	4	12	253 c
		0·858	52·0					144
Oct. 18						(192)	(876)	(991)

Mean Solar Time.	No. of Group, and Letter for Spot.	Distance from Centre in terms of Sun's Radius.	Position Angle from Sun's Axis.	Heliographic Longitude.	Latitude.
1881. 291ᵈ·022		0·790	290·3		
	589	0·240	39·4	288··	
	590	0·587	74·1	262··	
	591	0·618	62·7	262··	
	592	0·666	125·4	261··	
	592	0·739	123·0	257··	
		0·916	118·0		
Oct. 19		0·954	74·3		
291·978		0·854	293·7		
	589	0·198	342·4	289·	
	590	0·392	72·0	263·	
	590	0·410	66·5	262·	
	591	0·464	52·7	262·	
	591	0·505	55·1	259·	
	592	0·561	136·9	261·	
	592	0·583	133·7	259·	
	592	0·611	132·4	256·	
	593	0·943	70·3	214·	
Oct. 20		0·844	123·2		
292·775		0·907	297·1		
	589	0·315	300·3	291·	
	589	0·306	342·9	289·	
	590	0·229	57·9	263·	
	590	0·265	51·7	262·	
	591	0·346	36·8	262·	
	591	0·376	40·0	259·	
	592	0·466	151·4	261·	
	592	0·512	144·2	256·	
	593	0·843	99·2	218·	
	593	0·894	69·5	211·	
Oct. 21*		0·951	76·3		
293·784	589	0·487	294·6	288·	
	590	0·129	344·9	263·	
	590	0·102	352·4	262·	
	590	0·166	355·9	262·	
	591	0·276	358·8	262·	
	592	0·416	178·1	260·	
	592	0·427	172·7	258·	
	592	0·427	167·9	256·	
	594	0·296	16·7	253·	
	594	0·295	36·5	251·	
	594	0·305	38·4	250·	
	594	0·328	35·8	249·	
	593	0·716	66·1	217·	
	593	0·770	68·3	212·	
	593	0·789	67·5	210·	
Oct. 22*		0·895	74·9		

* Indian photo.   The Areas of Spots and Faculæ are expressed in Millionths of the Sun's visible Hemispher

MEASURES of POSITIONS and AREAS of SPOTS and FACULÆ upon the SUN'S DISK on PHOTOGRAPHS taken in the

Mean Solar Time.	No. of Group and Letter for Spot.	Distance from Centre in terms of Sun's Radius.	Position Angle from Sun's Axis.	HELIOGRAPHIC Longitude.	Latitude.	SPOTS. Area of UMBRA for each Spot (and for Day).	Area of WHOLE for each Spot (and for Day).	FACULÆ. Area for each Group (and for Day).
1881.			°	°	°			
294ᵈ·788	589	0·665	288·5	289·2	+15·9	63	337	
	590	0·289	296·4	263·7	+12·1	7	16	
	590	0·289	306·4	262·1	+14·7	20	123	
	591	0·357	321·7	262·0	+21·1	25	93	
	592	0·460	205·9	260·7	−19·4	41	133	
	592	0·437	197·7	256·5	−19·6	17	87	
	594	0·284	342·1	253·7	+20·6	3	16	
	594	0·249	344·4	252·4	+18·8	18	62	
	594	0·269	358·3	248·9	+20·6	38	196	
	593	0·622	63·3	212·2	+20·2	0	6	
	593	0·647	63·5	210·3	+20·8	0	14	
		0·798	74·8					1690
Oct. 23*						(232)	(1083)	(1690)
295·780	589	0·805	286·4	288·6	+16·1	53	357	
	590	0·491	287·1	263·9	+12·6	0	5	
	590	0·473	292·6	261·0	+14·8	18	107	
	591	0·511	304·4	262·1	+21·2	24	88	
	592	0·569	224·7	260·3	−19·3	21	144	
	592	0·528	220·2	256·4	−19·1	26	107	
	594	0·416	313·8	234·0	+21·4	0	9	
	594	0·393	311·9	253·3	+19·8	17	68	
	594	0·348	321·0	248·8	+20·4	50	249	
	593	0·406	47·5	216·7	+20·5	0	9	
	593	0·465	53·3	211·9	+20·6	0	9	
	593	0·496	54·4	209·8	+21·2	0	14	
		0·939	67·0					603
Oct. 24*						(210)	(1166)	(603)
296·930	589	0·924	285·2	287·8	+15·9	104	390	574 f
	591	0·693	295·9	261·9	+21·2	26	102	
	590	0·665	287·5	261·0	+15·1	44	142	
	592	0·727	236·3	260·0	−20·1	36	150	
	592	0·683	234·9	256·1	−19·2	27	68	
	594	0·588	299·0	753·3	+20·6	15	43	
	594	0·537	299·8	249·7	+19·6	9	95	
	594	0·523	303·2	247·9	+20·9	45	214	
	593	0·337	30·5	209·6	+21·6	5	10	
		0·837	68·7					250
		0·916	125·8					224
Oct. 25						(311)	(1214)	(1048)
297·906		0·790	310·3					67
	589	0·989	284·3	289·5	+14·9	0	207	639 n f
	591	0·834	291·8	262·9	+20·7	7	84	198 c
	590	0·814	284·7	261·3	+14·7	26	152	182 c
	592	0·845	242·7	260·0	−20·0	36	142	
	592	0·809	242·1	256·2	−19·1	19	58	438 c
	594	0·749	292·2	254·5	+19·7	2	12	
	594	0·691	292·0	249·6	+18·4	4	25	
	594	0·680	295·3	248·1	+20·5	30	212	368 c
	593	0·292	352·0	209·7	+21·5	0	6	
	595	0·973	110·7	133·4	−18·8	9	37	361 c
		0·753	61·7					230
		0·865	128·7					235
Oct. 26						(133)	(885)	(2718)

Mean Solar Time.	No. of Group and Letter for Spot.	Distance from Centre in terms of Sun's Radius.	Position Angle from Sun's Axis.	HELIOGRAPHIC Longitude.	Latitude.
1881.				°	°
298ᵈ·785	590	0·909	284·2	261·2	+14·9
	592	0·929	246·0	260·2	−20·2
	592	0·903	145·9	256·4	−19·4
	594	0·813	289·5	249·4	+18·6
	594	0·799	292·2	247·6	+20·4
	594	0·805	294·6	247·9	+22·5
	595a	0·206	22·4	141·5	+15·6
	595	0·912	113·7	133·7	−19·3
Oct. 27*					
299·775	590	0·977	284·4	260·7	+15·0
	592	0·981	248·7	258·8	−19·8
	594	0·902	291·4	246·5	+21·2
	595	0·811	118·1	133·4	−19·4
		0·876	118·3		
Oct. 28*					
300·986		0·928	245·0		
		0·905	299·0		
		0·783	292·7		
	594	0·990	291·5	249·3	+21·9
	595	0·657	126·9	132·9	−19·5
		0·680	125·7		
		0·872	75·5		
		0·898	96·7		
		0·956	119·3		
Oct. 29					
302·068		0·916	289·7		
	596	0·700	246·1	193·3	−13·2
	596	0·650	242·6	186·6	−13·9
		0·870	68·2		
Oct. 30					
302·759		0·888	290·8		
Oct. 31*					
303·653		0·878	252·1		
	596a	0·900	284·8	193·5	+15·1
	596a	0·875	285·4	192·3	+15·5
Nov. 1*					
304·760		0·963	183·9		
	598	0·968	76·4	40·9	+14·2
Nov. 2*					
305·789	598	0·887	76·2	40·9	+14·1
Nov. 3*					

* Indian photo. The Areas of Spots and Faculæ are expressed in Millionths of the Sun's visible Hemisphere.

MEASURES of POSITIONS and AREAS of SPOTS and FACULÆ upon the SUN'S DISK on PHOTOGRAPHS taken in the YEAR 1881—*continued.*

Mean Solar Time.	No. of Group, and Letter for Spot.	Distance from Centre in terms of Sun's Radius.	Position Angle from Sun's Axis.	Longitude.	Latitude.	Area of UMBRA for each Spot (and for Day).	Area of WHOLE for each Spot (and for Day).	Area for each Group (and for Day).	Area for each Group (and for Day).
1881. 306ᵈ·764		0·942	295·8					1276	
	598	0·767	74·9	40·9	+14·0	22	117	1123 f	
Nov. 4*						(22)	(117)	(2399)	
307·998		0·858	242·2					730	
	597	0·802	280·6	127·3	+10·7	2	7	177 c	
	598	0·571	70·5	40·6	+14·1	29	115		
		0·896	61·3					179	
Nov. 5						(31)	(122)	(1086)	
308·789		0·936	244·9					1221	
	598	0·415	62·8	41·4	+14·3	24	104		
		0·953	74·7					973	
Nov. 6*						(24)	(104)	(2194)	
309·793		0·960	245·0					856	
	598	0·242	38·3	41·6	+14·4	19	97		
	601a	0·514	45·8	26·8	+24·1	0	28		
	601a	0·563	47·4	23·3	+25·5	0	13		
	600	0·840	73·2	353·9	+16·0	0	38		
	600	0·879	75·5	342·3	+14·4	0	59	357 c	
	601	0·939	103·6	342·1	−11·5	0	104		
	601	0·977	103·8	334·1	−12·7	0	101	603 c	
Nov. 7*						(19)	(440)	(1816)	
310·764	598	0·208	340·1	41·9	+14·6	18	97		
	600	0·706	70·9	354·0	+15·8	0	44		
	600	0·760	73·1	349·1	+15·0	5	48		
	601	0·839	106·5	342·6	−11·8	0	72		
	601	0·912	105·8	333·7	−12·9	0	107	607 e	
Nov. 8*						(23)	(368)	(607)	
311·975		0·918	294·0					237	
	598	0·394	301·3	42·0	+14·8	24	112		
	599	0·318	26·1	13·2	+19·7	7	14		
	599	0·500	63·6	354·1	+15·7	11	27		
	600	0·559	64·3	352·1	+16·1	0	5		
	600	0·558	67·3	348·9	+15·3	10	47		
	601	0·669	111·5	342·4	−11·7	22	100		
	601	0·769	109·5	333·9	−12·6	15	64	150 e	
		0·904	65·0					67	
Nov. 9						(89)	(369)	(454)	
312·656	598	0·514	293·8	41·6	+14·5	20	110		
	599	0·275	355·2	14·1	+19·0	11	46		
	599	0·281	359·4	13·9	+19·4	7	23		
	599	0·265	8·2	10·4	+18·4	8	33		
	600	0·365	55·7	354·6	+14·9	0	17		
	600	0·450	61·8	348·6	+15·3	7	34		
	601	0·555	117·1	342·5	−11·9	26	91		
	601	0·667	114·1	334·1	−13·3	0	53		
		0·952	71·7					1397	
Nov. 10*						(79)	(407)	(1397)	
1881. 313ᵈ·784	598	0·704	287·5	41·5	+14·5	22	72		
	599	0·387	315·9	14·3	+19·0	34	143		
	599	0·337	326·9	9·0	+19·4	0	63		
	600	0·293	6·0	356·0	+20·0	0	15		
	600	0·255	24·7	351·5	+16·4	0	16		
	600	0·281	27·0	350·1	+17·5	0	13		
	600	0·274	36·6	348·1	+15·7	0	18		
	601	0·358	133·7	342·6	−11·3	24	92		
	601	0·499	124·2	332·8	−13·4	0	18		
	603	0·946	73·7	286·7	+16·5	109	381	1035 e	
Nov. 11*						(189)	(831)	(1095)	
314·940		0·906	254·3						106
	598	0·863	284·4	41·8	+13·9	24	97		
	599	0·579	299·3	14·8	+18·9	20	155		
	599	0·514	304·8	9·1	+19·6	9	37		
	600	0·263	333·3	349·7	+15·4	4	31		
	601	0·246	183·0	343·4	−11·2	17	66		
	601	0·310	151·6	334·0	−12·9	2	6		
	602	0·593	68·5	308·0	+14·9	5	21		
	602	0·646	70·9	349·3	+14·4	8	27		
	603	0·800	75·7	290·1	+13·1	6	30		
	603	0·833	72·3	287·1	+16·3	118	420	963 e	
Nov. 12						(219)	(890)	(1069)	
316·047	598	0·963	283·7	42·3	+13·9	20	77	139 nf	
	604	0·929	239·3	31·6	−27·1	23	132		
	604	0·903	234·5	26·0	−30·1	15	125	163 e	
	599	0·754	291·5	15·2	+18·6	13	162		
	599	0·722	294·8	13·1	+19·2	0	9		
	599	0·699	295·9	9·8	+19·8	13	27	412 e	
	601	0·358	228·5	343·8	−11·0	19	69		
	601	0·382	55·2	309·2	+15·2	7	13		
	602	0·455	62·4	303·5	+14·7	2	24		
	603	0·571	72·0	294·4	+12·5	20	65		
	603	0·611	71·5	291·6	+13·4	4	39		
	603	0·660	71·8	288·0	+14·0	15	34		
	603	0·678	69·3	287·1	+16·6	96	422		
	605	0·972	66·3	251·6	+23·7	19	57	218 f	
		0·776	62·0					472	
		0·879	68·5					222	
		0·926	112·2					157	
		0·955	128·2					102	
Nov. 13						(266)	(1255)	(1885)	
316·786		0·907	297·9						275
		0·889	287·6						372
	604	0·972	242·1	31·6	−26·3	26	111		
	604	0·946	236·3	24·2	−30·5	19	71	598 e	
	599	0·857	289·8	16·3	+18·3	0	45		
	599	0·840	290·6	14·3	+18·7	0	29		
	599	0·793	293·2	9·0	+19·9	0	25	524 e	
	601	0·483	240·7	343·7	−11·2	16	36		
	603	0·428	64·7	295·1	+13·0	0	69		

* Indian photo.   The Areas of Spots and Faculæ are expressed in Millionths of the Sun's visible Hemisphere.

MEASURES of POSITIONS and AREAS of SPOTS and FACULÆ upon the Sun's DISK on PHOTOGRAPHS taken in the YEAR 1881—*continued.*

Mean Solar Time.	No. of Group, and Letter for Spot.	Distance from Centre in terms of Sun's Radius.	Position Angle from Sun's Axis.	HELIOGRAPHIC Longitude.	HELIOGRAPHIC Latitude.	SPOTS Area of UMBRA for each Spot (and for Day).	SPOTS Area of WHOLE Spot (and for Day).	FACULÆ Area for each Group (and for Day).
1881.								
316d·786	603	0·483	67·6	291·2	+13·0	5	50	
	603	0·518	68·5	288·8	+13·3	4	27	
	603	0·557	64·2	287·0	+16·3	53	356	
	605	0·929	65·8	251·1	+23·5	0	58	
Nov. 14*						(123)	(877)	(1769)
317·795		0·920	289·6					788
	603	0·245	35·1	296·7	+14·1	8	37	
	603	0·247	42·4	295·2	+13·0	0	48	
	603	0·291	51·5	291·5	+12·9	0	49	
	603	0·323	55·2	289·3	+13·1	0	18	
	603	0·383	50·8	287·0	+16·5	51	338	
	603	0·360	67·2	285·4	+10·4	0	8	
	603	0·386	57·9	285·4	+14·3	0	8	
	605	0·831	63·9	251·1	+22·9	0	53	
	607	0·972	79·1	228·6	+11·2	0	130	
	606	0·986	73·9	224·6	+16·3	0	91	1414 c
Nov. 15*						(59)	(780)	(2202)
318·635		0·974	188·3					676
		0·803	185·3					424
	603	0·211	344·5	297·1	+14·1	32	80	
	603	0·184	11·5	291·8	+12·8	16	54	
	603	0·272	24·8	287·1	+16·6	55	341	
	605	0·721	60·0	251·4	+22·8	11	45	
	606	0·881	70·6	233·1	+18·2	33	108	327 c
	606	0·938	73·5	226·2	+16·2	0	42	832 c
	607	0·910	71·2	228·6	+10·8	55	224	
	607	0·976	79·5	216·5	+10·7	21	211	1428 c
Nov. 16*						(223)	(1105)	(3687)
319·967		0·971	185·2					585
	601	0·931	156·0	343·6	-12·1	0	9	483 f
	603	0·424	299·7	198·6	+14·2	34	113	
	603	0·377	303·3	295·2	+14·1	0	3	
	603	0·312	304·9	291·9	+12·7	4	34	
	603	0·308	325·5	286·8	+16·9	75	326	
	605	0·530	47·9	251·2	+22·9	14	42	
	606	0·672	66·9	236·3	+17·0	34	191	
	606	0·713	67·6	232·9	+17·4	14	61	
	606	0·749	66·3	230·1	+19·1	33	153	
	606	0·780	70·7	126·5	+16·4	5	13	216 c
	607	0·750	77·2	228·4	+11·1	48	279	
	607	0·803	79·2	223·3	+10·0	10	41	
	607	0·833	82·7	220·2	+ 7·4	4	17	
	607	0·865	78·0	216·9	+11·5	85	441	398 c
		0·755	125·3					191
		0·822	53·5					234
		0·865	71·0					414
Nov. 17						(360)	(1723)	(2531)
320·972		0·973	256·2					363
		0·884	190·5					22
		0·877	203·3					50
		0·778	197·7					213
	603	0·606	190·4	298·8	+13·9	20	99	

Mean Solar Time.	No. of Group, and Letter for Spot.	Distance from Centre in terms of Sun's Radius.	Position Angle from Sun's Axis.	HELIOGRAPHIC Longitude.	HELIOGRAPHIC Latitude.	SPOTS Area of UMBRA for each Spot (and for Day).	SPOTS Area of WHOLE Spot (and for Day).	FACULÆ Area for each Group (and for Day).
1881.								
320d·972	603	0·514	202·1	292·3	+13·0	0	7	
	603	0·462	304·1	286·6	+17·0	71	326	
	605	0·410	28·7	250·8	+23·1	5	21	
	606	0·503	58·0	236·7	+17·4	32	168	
	606	0·533	59·0	234·9	+17·7	8	12	
	606	0·604	60·8	229·4	+18·9	30	116	
	606	0 628	66·4	226·4	+16·3	0	16	444 f
	607	0·575	73·2	229·1	+11·4	70	275	
	607	0·694	76·5	219·9	+10·9	7	40	
	607	0·691	81·6	219·7	+ 7·4	0	8	
	607	0·735	76·2	216·5	+11·6	133	488	317 c
		0·878	67·0					480
		0·924	105·3					315
Nov. 18						(376)	(1576)	(2204)
322·001		0·896	292·2					222
		0·715	318·2					109
	603	0·778	285·8	299·7	+13·5	27	106	90 c
	603	0·696	287·1	291·5	+13·3	0	6	295 c
	603	0·640	294·6	286·9	+17·1	60	278	148 n p
	606	0·339	36·6	237·4	+17·7	43	172	
	606	0·363	41·6	235·0	+17·7	0	16	
	606	0·433	48·1	229·8	+18·7	13	60	
	606	0·445	56·1	227·1	+16·2	0	9	
	607	0·389	64·4	228·7	+11·6	48	227	
	607	0·555	72·5	217·1	+11·4	102	442	
		0·842	58·5					359
		0·845	108·8					147
Nov. 19						(293)	(1316)	(1370)
323·036	603	0·903	283·6	300·0	+13·1	49	121	379 n f
	603	0·791	290·1	286·6	+17·0	67	343	745 n
	606	0·777	354·0	237·6	+17·9	31	135	
	606	0·302	17·8	230·4	+18·6	9	26	
	607	0·286	31·8	226·9	+15·9	0	3	
	607	0·105	37·9	228·6	+11·2	64	251	
	607	0·269	62·6	222·0	+ 9·0	4	16	
	607	0·357	62·3	217·2	+11·4	119	479	
	609	0·843	108·2	180·5	-14·1	8	24	118 s
Nov. 20						(351)	(1398)	(1242)
323·950	603	0·972	282·6	300·1	+12·7	19	92	
	603	0·884	287·9	286·3	+16·8	54	265	684 p
	606	0·356	310·3	237·6	+17·6	17	125	
	606	0·291	340·5	229·7	+17·7	7	31	
	607	0·181	331·0	229·0	+10·9	25	155	
	607	0·177	345·1	226·5	+11·6	3	39	
	607	0·140	0·1	223·9	+ 9·9	30	73	
	607	0·103	16·7	222·2	+ 7·5	6	26	
	607	0·187	30·9	218·3	+11·0	150	832	
	610	0·348	44·5	209·2	+16·1	0	3	
	611	0·526	48·6	198·8	+22·0	0	13	
	609	0·731	113·1	179·9	-15·3	0	15	
		0·942	110·3					126
		0·943	63·2					206
Nov. 21						(311)	(166g)	(1016)

* Indian photo.   The Areas of Spots and Faculæ are expressed in Millionths of the Sun's visible Hemisphere.   † Melbourne photo.

MEASURES of POSITIONS and AREAS of SPOTS and FACULÆ upon the SUN's DISK on PHOTOGRAPHS taken in the YEAR 1881—*continued*.

Mean Solar Time	No. of Group, and Letter for Spot	Distance from Centre in terms of Sun's Radius	Position Angle from Sun's Axis	HELIOGRAPHIC Longitude	HELIOGRAPHIC Latitude	SPOTS Area of UMBRA for each Spot (and for Day)	SPOTS Area of WHOLE for each Spot (and for Day)	FACULÆ Area for each Group (and for Day)
1881. 324ᵈ·793	603	0·962	287·2	286·5	+17·0	51	186	1354 c
	606	0·487	304·4	137·6	+17·5	18	59	
	606	0·382	317·6	128·9	+18·0	7	47	
	607	0·315	199·3	229·0	+10·5	31	140	
	607	0·260	305·4	225·2	+10·3	16	81	
	607	0·191	300·0	222·3	+ 7·1	5	17	
	607	0·192	326·8	218·9	+10·9	103	756	
	607	0·220	335·8	217·3	+13·5	6	32	
	609	0·603	119·4	179·9	−15·7	4	18	
Nov. 22*						(241)	(1336)	(1354)
326·004		0·899	293·8					225
		0·898	238·8					92
	606	0·690	293·9	238·0	+17·4	17	72	
	606	0·602	296·6	230·9	+16·9	9	111	
	607	0·526	287·3	227·4	+10·3	28	248	
	607	0·447	283·8	222·6	+ 7·5	4	7	
	607	0·412	294·9	219·1	+11·4	171	972	
	612	0·203	4·2	195·9	+13·2	0	6	
	609	0·391	139·9	181·7	−15·8	17	95	
		0·854	78·2					116
		0·922	113·7					302
Nov. 23						(246)	(1511)	(735)
326·784		0·877	296·1					645
	606	0·796	290·9	237·5	+17·4	13	41	
	606	0·752	291·0	233·4	+16·6	16	94	
	606	0·687	293·6	227·5	+17·1	9	21	
	607	0·683	284·2	228·7	+10·7	21	123	
	607	0·650	282·8	226·3	+ 9·4	12	29	
	607	0·623	186·6	223·9	+11·4	0	38	
	607	0·542	287·4	218·1	+10·6	108	742	
	609	0·294	174·3	184·8	−15·4	3	33	
	609	0·309	165·4	181·9	−15·9	0	10	
	609	0·312	159·0	179·6	−15·9	7	41	
Nov. 24*						(189)	(1172)	(645)
327·798		0·821	281·7					544
	606	0·913	288·4	238·2	+17·4	0	76	
	606	0·889	288·3	234·8	+16·8	17	196	865 c
	606	0·854	280·7	228·2	+17·1	11	112	
	607	0·836	281·8	229·3	+10·6	17	57	
	607	0·814	280·4	227·2	+ 9·2	0	17	
	607	0·798	282·6	225·5	+10·9	0	77	
	607	0·717	283·4	218·2	+10·6	145	558	
	609	0·316	200·5	180·0	−16·3	0	54	
Nov. 25*						(190)	(1157)	(1409)
328·938		0·903	284·2					853
		0·875	67·9					200
	606	0·977	286·7	235·4	+16·6	101	580	
	606	0·953	288·1	229·8	+17·6	15	112	
	607	0·947	280·1	229·1	+ 9·9	20	78	
	607	0·873	280·9	218·4	+10·1	177	565	
	609	0·467	229·5	179·7	−16·5	5	39	
Nov. 26						(318)	(1374)	(1053)
1881. 330ᵈ·050		0·911	306·0					141
		0·848	253·7					118
	607	0·971	179·9	219·3	+ 9·8	134	693	756 n f
	609	0·582	247·8	183·9	−14·1	2	14	198 p
	613	0·965	110·1	70·3	−19·0	8	20	
	614	0·992	109·9	61·7	−19·6	16	322	401 e
		0·649	62·8					112
		0·866	62·0					105
Nov. 27						(160)	(1049)	(1832)
330·954		0·930	255·8					93
	615	0·929	255·0	198·7	−13·5	0	10	
	609	0·805	251·8	183·4	−13·9	5	16	379 e
	616	0·747	56·5	88·4	+25·0	5	18	
	616	0·777	58·8	84·9	+24·3	3	16	97 e
	613	0·891	112·2	70·9	−19·1	16	48	
	614	0·955	111·1	60·6	−19·9	143	601	482 e
Nov. 28						(172)	(709)	(1051)
331·995		0·909	252·0					400
	617	0·535	70·8	87·0	+10·8	5	12	
	616	0·597	45·8	89·7	+25·2	10	20	
	616	0·639	51·1	84·9	+24·2	2	5	
	613	0·763	116·1	71·6	−19·0	13	53	
	613	0·790	114·1	68·7	−18·3	4	18	
	614	0·860	114·0	61·3	−20·0	87	711	
	614	0·908	109·6	54·4	−17·4	6	19	751 e
		0·820	62·3					53
Nov. 29						(127)	(838)	(1204)
332·608	614	0·788	115·5	61·0	−19·3	151	428	
Nov. 30†						(151)	(428)	
333·783	614	0·629	123·3	60·5	−19·7	86	491	
	619	0·656	64·9	23·0	+24·1	0	94	812 e
		0·956	122·7					996
Dec. 1*						(86)	(585)	(1808)
334·904		0·940	300·7					171
	618	0·813	246·0	131·1	−19·0	44	76	
	618	0·778	245·4	127·6	−18·5	26	59	352 e
	614	0·452	142·8	62·6	−20·6	27	100	
	614	0·466	136·7	59·8	−19·3	143	419	
	619	0·839	60·2	26·3	+24·9	54	157	
	619	0·867	59·5	23·3	+26·3	25	122	
	619	0·863	62·3	23·0	+23·9	37	133	
	619	0·900	60·5	18·7	+26·5	33	153	
	619	0·931	61·8	13·6	+26·3	0	14	554 e
		0·888	125·3					359
		0·934	70·7					207
Dec. 2						(389)	(1333)	(1643)

* Indian photo.　　The Areas of Spots and Faculæ are expressed in Millionths of the Sun's visible Hemisphere.　　† Melbourne photo.

MEASURES of POSITIONS and AREAS of SPOTS and FACULÆ upon the SUN's DISK on PHOTOGRAPHS taken in the YEAR 1881—*continued.*

Mean Solar Time.	No. of Group and Letter for Spot.	Distance from Centre in terms of Sun's Radius.	Position Angle from Sun's Axis.	Heliographic Longitude.	Heliographic Latitude.	Area of UMBRA for each Spot (and for Day).	Area of WHOLE Spot (and for Day).	Area for each Group (and for Day).
1881.				°	°			
335d·976	618	0·916	248·3	129·8	−19·7	98	512	525 c
	613	0·349	194·7	70·8	−19·3	0	7	
	614	0·364	173·6	62·9	−20·8	0	20	
	614	0·355	164·3	59·6	−19·6	103	331	
	614	0·413	143·7	50·4	−19·0	0	24	
	619	0·708	·44	26·3	+24·5	89	557	
	619	0·768	56·2	20·6	+25·4	25	239	601 e
		0·704	132·0					329
		0·892	70·8					330
Dec. 3						(315)	(1690)	(1785)
336·666	618	0·976	249·5	133·5	−19·9	0	282	
	618	0·954	249·9	127·4	−19·0	5	195	1178 e
	620	0·470	320·1	75·1	+21·2	0	35	
	614	0·340	188·0	59·2	−19·4	96	332	
	619	0·611	46·9	27·0	+24·8	75	502	
	619	0·634	52·7	23·3	+22·7	0	10	
	619	0·669	49·0	22·2	+26·1	10	69	
	619	0·694	52·1	19·1	+25·4	15	82	
	619	0·720	54·2	16·3	+25·0	0	18	261 f
		0·901	68·7					802
		0·960	104·4					603
Dec. 4*						(191)	(1525)	(2844)
337·780		0·940	286·9					904
	620	0·641	307·6	76·6	+20·2	0	20	
	620	0·602	307·3	72·4	+21·4	0	29	
	614	0·438	220·4	59·0	−19·3	104	269	
	619	0·475	28·3	27·3	+24·7	85	531	
	619	0·525	3·47	22·3	+25·6	4	55	
	619	0·508	37·6	21·9	+23·7	0	18	
	619	0·564	39·9	18·0	+25·6	9	151	
		0·870	106·7					864
Dec. 5*						(202)	(1074)	(1768)
338d·930	620	0·762	208·1	72·4	+20·9	5	10	297 n p
	614	0·637	235·3	60·6	−21·3	7	38	
	614	0·609	237·9	59·4	−18·9	60	156	620 n p
	619	0·426	356·6	28·1	+25·0	123	590	
	619	0·450	14·2	19·5	+25·7	29	192	
		0·950	63·2					143
		0·973	107·0					216
Dec. 6						(223)	(986)	(1376)
339·802		0·915	295·7					421
		0·832	246·2					559
	614	0·771	242·3	61·8	−21·1	0	24	
	614	0·737	244·3	59·3	−18·7	69	219	
	619	0·475	333·3	28·6	+24·9	94	425	
	619	0·456	339·7	25·0	+25·0	18	114	
	619	0·446	347·6	21·1	+15·5	23	122	
	619	0·446	357·6	16·2	+26·2	0	13	
		0·905	108·9					364
		0·930	61·2					567
Dec. 7*						(204)	(917)	(1911)
1881.				°	°			
340d·780	614	0·858	248·2	59·1	−18·7	60	271	
	619	0·615	316·7	30·7	+26·7	0	20	
	619	0·595	315·1	29·5	+24·5	39	233	
	619	0·569	318·5	20·6	+24·9	38	190	
	619	0·520	326·9	20·4	+25·5	27	146	
Dec. 8*						(164)	(860)	
341·791	614	0·950	250·2	59·4	−18·9	44	138	763 n f
	619	0·721	306·0	28·6	+24·7	63	247	
	619	0·690	308·6	25·2	+25·1	0	54	
	619	0·642	313·0	20·0	+25·5	0	67	
	625	0·924	71·8	282·7	+16·6	16	73	566 e
	621	0·978	61·8	272·7	+27·5	0	66	805 e
Dec. 9*						(123)	(645)	(2134)
342·831		0·963	249·5					880
	619	0·845	299·7	28·5	+24·3	38	145	
	619	0·817	300·9	15·1	+24·3	3	27	
	619	0·782	304·5	10·5	+25·8	0	25	
	619	0·741	306·7	16·1	+25·8	0	19	
	619	0·711	309·6	12·6	+26·4	0	11	881 e
	622	0·391	316·0	351·3	+15·7	8	25	
	622	0·351	324·0	347·3	+15·9	0	15	
	625	0·818	69·1	282·3	+16·6	11	40	297 e
	621	0·917	59·7	272·2	+27·3	8	128	697 e
		0·974	104·9					744
Dec. 10*						(68)	(435)	(3499)
343·728		0·861	234·3					610
	619	0·926	296·1	28·2	+23·7	41	143	1454 f
	622	0·543	328·8	352·5	+14·6	9	22	
	622	0·467	298·7	347·9	+12·3	0	18	
	622	0·434	303·8	344·9	+13·3	0	19	
	625	0·694	65·0	282·4	+16·5	13	50	
	621	0·838	56·6	271·6	+27·0	14	71	835 f
	627a	0·881	106·0	262·5	−14·4	0	33	490
	627	0·973	63·0	248·7	+26·0	13	166	889 f
Dec. 11*						(90)	(522)	(4278)
344d·780	619	0·985	294·2	28·1	+23·7	35	114	2420 n f
	622	0·720	289·9	353·4	+13·5	0	25	
	622	0·610	292·8	344·5	+13·0	0	16	
	625	0·534	56·0	282·0	+16·6	13	37	
	621	0·717	50·5	271·4	+26·4	11	51	
	627a	0·746	109·2	262·8	−14·7	0	12	804 f
	627	0·902	60·0	248·9	+26·4	100	358	
	627	0·942	60·7	242·3	+27·1	18	51	1030 e
	628	0·961	72·0	236·6	+17·1	22	111	{ 1351 e / 1276 n f
Dec. 12*						(199)	(775)	(6882)
345·781		0·985	297·0					1980
	629n	0·967	287·6	10·3	+16·8	0	48	816
	625	0·380	38·5	282·0	+16·3	3	25	
	621	0·600	40·0	270·8	+26·4	10	84	

* Indian photo.　　The Areas of Spots and Faculæ are expressed in Millionths of the Sun's visible Hemisphere.

MEASURES of POSITIONS and AREAS of SPOTS and FACULÆ upon the Sun's Disk on PHOTOGRAPHS taken in the YEAR 1881—*continued.*

Mean Solar Time.	No. of Group and Letter for Spot.	Distance from Centre in terms of Sun's Radius.	Position Angle from Sun's Axis.	HELIOGRAPHIC Longitude.	Latitude.	SPOTS Area of UMBRA for each Spot (and for Day).	Area of WHOLE Spot (and for Day).	FACULÆ Area for each Group (and for Day).
1881. 345ᵈ·781	627	0·798	54·9	249·5	+26·6	64	399	
	627	0·855	57·7	242·5	+26·6	0	180	
	628	0·870	70·3	237·7	+16·6	31	110	1695 *f*
	629	0·962	79·1	222·7	+10·3	11	79	
Dec. 13*	629	0·979	78·5	218·6	+11·1	0	75	961 *c*
						(119)	(1000)	(5452)
346·954	622	0·949	285·0	351·3	+13·8	0	24	258 *c*
	623	0·607	295·7	315·0	+14·3	20	79	
	624	0·286	214·5	290·3	−14·6	0	4	
	625	0·307	357·4	281·5	+16·7	8	32	
	616	0·391	41·6	265·1	+15·9	5	15	
	621	0·491	20·3	269·0	+26·2	13	49	
	627	0·670	45·3	248·5	+27·1	86	414	
	627	0·738	50·2	241·3	+27·2	12	158	537 *f*
	628	0·721	63·2	237·8	+16·8	21	126	294 *f*
	629	0·855	77·1	223·7	+10·4	23	97	
	629	0·892	76·3	218·8	+11·7	0	57	
	629	0·925	76·0	214·1	+12·5	0	32	406 *c*
Dec. 14						(188)	(1087)	(1495)
347·782		0·984	285·2					1156
	623	0·751	290·1	316·3	+14·1	9	129	
	625	0·365	327·0	281·7	+16·6	0	20	
	621	0·460	359·3	272·0	+26·1	2	32	
	626	0·280	9·5	267·1	+14·8	3	24	
	627	0·581	34·3	248·2	+27·5	54	363	
	627	0·63g	43·1	241·1	+27·1	12	87	
	618	0·58g	59·4	238·0	+16·4	27	131	
	629	0·71g	71·3	225·8	+12·4	0	22	
	629	0·736	69·3	224·7	+14·2	0	11	
	629	0·743	75·2	213·1	+10·1	16	42	
	629	0·792	74·9	218·8	+11·1	0	23	
Dec. 15*	629	0·842	73·8	214·0	+12·9	12	57	741 *c*
						(135)	(941)	(1897)
348·849	623	0·907	287·0	319·4	+14·8	0	147	320 *p*
	625	0·513	306·0	281·3	+16·3	0	16	
	621	0·507	335·6	269·2	+26·2	0	36	
	626	0·335	323·8	267·5	+14·4	0	25	
	627	0·505	15·4	247·1	+27·7	72	362	
	627	0·53g	26·6	240·1	+27·5	7	65	
	628	0·414	44·0	238·0	+16·4	31	126	
	629	0·530	63·8	226·8	+12·4	0	17	
	629	0·577	62·3	224·1	+14·4	0	13	
	629	0·56g	70·3	223·0	+9·9	6	28	
	629	0·63o	70·5	218·7	+11·1	0	10	
Dec. 16*	629	0·696	69·7	213·9	+13·0	9	22	(320)
						(125)	(867)	
349·958		0·936	296·3					206
	623	0·982	285·2	319·2	+14·6	0	80	244 *f*
	627	0·502	36·1	246·8	+28·1	84	386	
	627	0·493	2·5	239·8	+27·9	0	35	
	628	0·310	8·1	238·6	+16·3	17	83	
Dec. 17	629	0·361	56·1	223·1	+10·2	11	31	(450)
						(112)	(615)	

Mean Solar Time.	No. of Group and Letter for Spot.	Distance from Centre in terms of Sun's Radius.	Position Angle from Sun's Axis.	HELIOGRAPHIC Longitude.	Latitude.	SPOTS Area of UMBRA for each Spot (and for Day).	Area of WHOLE Spot (and for Day).	FACULÆ Area for each Group (and for Day).
1881. 350ᵈ·975								840
	627	0·849	299·7					
	627	0·572	329·7	246·8	+27·9	108	424	
	627	0·528	340·2	239·4	+28·1	3	10	
	628	0·358	328·9	238·8	+16·2	22	110	
	629	0·261	9·3	225·3	+13·3	8	29	
	629	c·216	17·9	223·9	+10·2	3	15	
		0·929	62·7					119
Dec. 18						(144)	(588)	(959)
352·004		0·908	300·2					662
		0·817	290·3					245
		0·781	250·8					168
	627	0·683	316·3	246·4	+28·0	75	365	
	628	0·506	307·0	239·0	+16·1	29	96	
	629	0·321	328·1	234·3	+14·1	0	23	
	629	0·247	329·2	221·6	+10·5	0	8	
		0·873	57·8					219
		0·949	72·0					76
		0·976	108·0					503
Dec. 19						(104)	(492)	(1873)
353·054		0·940	286·2					83
		0·931	303·2					418
	627	0·805	307·3	246·6	+27·8	89	360	
	627	0·775	308·3	243·4	+27·2	0	37	
	628	0·674	296·5	239·0	+15·9	10	46	
	630	0·917	111·2	135·0	−20·2	0	21	
Dec. 20	630	0·949	109·7	129·5	−19·2	0	13	907 *n p*
						(99)	(477)	(1408)
353·950	627	0·892	302·9	246·3	+27·9	134	346	
	627	0·882	305·2	244·2	+29·4	20	130	469 *n f*
	628	0·804	291·9	239·4	+16·1	33	49	884 *f*
	630	0·815	114·2	135·2	−20·7	3	14	874 *c*
Dec. 21	630	0·870	111·4	129·6	−19·4	4	17	
						(194)	(556)	(2227)
354·778	627	0·950	300·7	245·4	+28·2	55	377	
	628	0·893	289·3	238·8	+16·1	0	16	363 *f*
	631	0·412	207·8	189·7	−23·3	0	12	
	631	0·436	198·1	186·4	−26·5	2	19	
	632	0·512	50·4	153·4	+17·1	2	9	
	632	0·547	55·4	149·9	+16·7	1	7	
	630	0·703	118·4	136·4	−21·1	1	9	
Dec. 22*	630	0·774	113·9	129·2	−19·6	0	13	237 *n f*
						(61)	(462)	(600)
355·686	627	0·910	287·0					1145
	627	0·988	299·5	244·1	+28·7	10	365	547 *f*
	631	0·533	226·8	190·6	−23·3	0	3	
	631	0·521	215·1	185·3	−27·2	2	6	
	632	0·392	36·9	151·6	+16·1	0	8	
Dec. 23*	633	0·368	51·0	148·8	+11·2	4	29	(1692)
						(16)	(411)	

* Indian photo.   The Areas of Spots and Faculæ are expressed in Millionths of the Sun's visible Hemisphere.

MEASURES of POSITIONS and AREAS of SPOTS and FACULÆ upon the SUN's DISK on PHOTOGRAPHS taken in the YEAR 1881—*continued.*

Mean Solar Time	No. of Group and Letter for Spot	Distance from Centre in terms of Sun's Radius	Position Angle from Sun's Axis	Heliographic Longitude	Heliographic Latitude	Area of UMBRA for each Spot (and for Day)	Area of WHOLE Spot (and for Day)	Faculæ Area for each Group (and for Day)
1881.								
356d.757	/	0.953	283.5					711
		0.989	69.8					253
		0.992	109.4					354
Dec. 24*						(0)	(0)	(1318)
357.684		0.964	283.4					286
	631	0.804	142.1	189.9	−23.6	6	38	258 c
	631	0.767	237.4	185.1	−26.1	1	7	121 c
		0.931	64.2					457
		0.975	112.1					689
Dec. 25*						(7)	(45)	(1811)
358.700	631	0.904	244.8	189.1	−23.8	0	43	407 e
	631	0.881	241.6	185.4	−26.0	0	29	
	634	0.490	147.6	109.0	−26.8	11	26	
	634	0.527	144.6	105.9	−27.7	7	15	
		0.863	63.3					236
		0.915	113.6					569
Dec. 26*						(18)	(113)	(1416)
359.690		0.964	145.9					815
		0.920	188.9					105
	634	0.406	172.3	109.5	−26.3	18	93	
	634	0.448	165.1	105.5	−28.2	19	73	
		0.784	58.9					144
Dec. 27*						(37)	(166)	(1064)

Mean Solar Time	No. of Group and Letter for Spot	Distance from Centre in terms of Sun's Radius	Position Angle from Sun's Axis	Heliographic Longitude	Heliographic Latitude	Area of UMBRA for each Spot (and for Day)	Area of WHOLE Spot (and for Day)	Faculæ Area for each Group (and for Day)
1881.								
360d.705		0.817	292.0					220
	635	0.986	287.9	178.4	+17.1	0	31	187 s
	634	0.419	202.5	109.8	−25.4	16	97	108 c
	634	0.436	189.2	104.1	−28.2	9	81	
		0.773	118.0					672
		0.968	61.7					974
Dec. 28*						(24)	(209)	(2161)
361.618	634	0.517	221.8	109.9	−25.2	25	73	
	634	0.503	208.7	103.5	−28.8	0	32	
	634	0.488	203.8	103.4	−27.7	7	38	
		0.937	71.9					770
		0.938	60.0					1554
Dec. 29†						(32)	(143)	(2324)
362.666		0.873	249.3					700
	634	0.663	233.7	109.9	−25.5	17	103	
	634	0.610	224.9	103.0	−28.2	13	25	
	636	0.850	57.3	21.6	+25.5	0	11	1163
Dec. 30*						(30)	(139)	(1863)

* Indian photo.   The Area of Spots and Faculæ are expressed in Millionths of the Sun's visible Hemisphere.   † Melbourne photo.

The following is a record of the absence of spots during the years 1878-81. Where the date is given without further remark, the photograph was taken at Greenwich. I. denotes the Indian, and Mn. the Mauritius record. When there was more than one record for the same day, preference has been given in the order named.

1878—

January 1 Mn., 2 Mn., 3, 4 Mn., 5 Mn., 7, 8 Mn., 9, 10, 11, 12 Mn., 13 Mn., 16 Mn., 17, 18, 19 Mn., 20 Mn., 21 Mn., 22 Mn., 29, 30, 31.

February 1, 2 Mn., 9 Mn., 10 Mn., 11 I., 12, 13 I., 14 I., 15 I., 16, 17 I., 18, 19 I., 20 I., 21, 22 I., 23, 24 I., 25 I., 28 I.

March 1, 2, 3 I., 7 I., 8 I., 10 Mn., 19 I., 20 I., 21 Mn., 22 Mn., 23 I., 24 I., 25, 26 I., 27, 28 I., 29 I., 30 I.

April 1, 2, 3 I., 4, 6, 7 I., 8, 9, 10 I., 11 I., 12, 13, 14 I., 15 I., 16 I., 17 I., 18, 19 I., 20 I., 21 I., 22 I., 23 I., 24 I., 25, 26 I., 27, 28 Mn., 29 Mn., 30.

May 1, 4, 5 Mn., 6, 7 Mn., 8 Mn., 9, 10, 11 Mn., 12 Mn., 13, 14 Mn., 15 Mn., 16, 17, 18, 19 Mn., 20 Mn., 21 Mn., 22 Mn., 23 I., 24 I., 25.

June 7 I., 8 I., 9 I., 11 I., 12, 13, 14 I., 15 I., 16 Mn., 17 I., 18, 19 I., 20, 21, 22, 23 Mn., 24 I., 25, 30 Mn.

July 1 I., 2 I., 3 I., 4, 5, 6, 8, 9, 10 I., 11, 12 I., 13, 15 I., 16, 17, 18, 19, 20, 21 I., 22, 23, 24 I., 25, 27, 28 I., 29 I., 30, 31 I.

August 1, 2, 3 I., 4 I., 5, 6, 7, 8, 9, 10 I., 11 I., 12, 13 I., 14 15, 16, 17, 19, 21 I., 22, 23 I., 24 I. 25 I., 26, 27 I., 28, 29, 30, 31 I.

September 1 I., 14, 15 I., 16, 17 I., 18 I., 19, 20, 21 I., 22 I., 23 I., 24 I., 25 I., 26, 27, 28 I., 29 I., 30.

1878—continued.

October 1, 2, 3, 4, 5, 6 I., 7, 8, 9, 10, 11, 12 I., 13 I., 14, 15 I., 16 I., 17 I., 18 I., 19, 20 I., 21, 22 I., 23, 24 I., 25, 26, 27 I., 28.

November 10 Ma., 11, 12, 13 Ma., 14 Ma., 15 Ma., 16, 17 Ma., 18, 19, 23, 24 I., 25 I., 26 I., 27 I., 28 I., 29, 30 I.

December 1 I., 2 I., 3 I., 4 I., 5 I., 6 I., 7 I., 8 I., 9 I., 10 I., 11 I., 12 I., 13, 14 I., 15 I., 16 I., 17 I., 18 I., 21 I., 22 I., 23 I., 24 I., 25 I., 26 I., 27 I., 28 I., 29 I., 30, 31 I.

1879—

January 1 I., 2 I., 3 I., 4 I., 5 I., 6 I., 7 I., 8, 9 I., 10, 11 Ma. I., 13 I., 14 I., 15 I., 16 I., 17 I., 18 I., 19 I., 20 I., 21 I., 22 I., 23 I., 24 I., 25 I., 26 I., 27 I., 28 I., 29 I.

February 1 I., 2 I., 3 I., 4 I., 5 I., 6 I., 7 I., 8, 9 I., 10 I., 11 I., 12, 13 Ma., 14 I., 17 I., 18 I., 19, 20, 21, 22 I., 23 I., 24, 25, 26 I., 27 Ma., 28 I.

March 1 I., 2 Ma., 3, 4, 5 I., 6 I., 7 I., 8, 9 I., 10, 11, 12 I., 13, 14 I., 15, 17 I., 18, 19, 20 I., 21, 22 I., 23 I., 24 I., 25 I., 26 I., 27 I., 28 I., 29, 30 I., 31.

April 1 Ma., 2, 3 Ma., 4, 5 Ma., 6 Ma., 7, 8, 9 Ma., 10 Ma., 11 Ma., 25, 26, 27 Ma. 28 Ma., 29, 30.

May 1, 2 Ma., 3, 4 Ma., 5, 6 Ma., 7 Ma., 13, 14, 15 Ma., 16 Ma., 17 I., 18 Ma., 19, 21, 22, 23 Ma., 24, 25 Ma., 26 Ma., 27, Ma., 28, 29, 30 Ma.

June 1 Ma., 2 Ma., 3, 4, 6 Ma., 7 Ma., 9, 10 Ma., 11, 12 Ma., 13 Ma., 14, 16 Ma., 17, 18, 19 Ma., 20, 21 Ma., 22 Ma., 23 Ma., 24, 25.

July 7, 8 Ma., 9, 10 Ma., 18, 21, Ma., 22 Ma., 23 Ma., 24, 25, 26 Ma., 28 Ma., 29, 30, 31 Ma.

August 1, 2, 4 Ma., 5 Ma., 6, 7, 8 Ma., 9, 14 Ma., 15, 16 Ma., 18, 20 Ma., 21 Ma., 22, 25, 26 Ma.

September 8, 9 Ma., 10, 11 Ma., 13 14, Ma., 15, 16, 17 Ma., 18 Ma., 19 Ma., 20 Ma., 21 Ma., 22, 23 Ma., 30.

October 4, 5 Ma., 6, 23 Ma., 24 Ma., 25, 26 Ma., 27, 28 Ma., 29 Ma., 30 Ma. , 31 Ma.

November 1 Ma., 2 Ma., 3 Ma., 19, 21 Ma., 22 Ma., 23 Ma., 24 Ma., 26 Ma.

December 5 I., 6, 7 Ma., 8 I., 9 I., 10 I., 11, 12 Ma., 13 Ma., 14 Ma., 15 Ma., 24, 26 Ma., 27 Ma., 28, 29 I., 30 I., 31 I.

1880—

January 1 I., 2 I., 22, 23 I., 25 I., 27, 28, 29 I.

February 15 I., 16 I., 17, 18, 19 I., 20.

March 5, 20, 21 I., 22 I., 23 I., 24, 25, 26 I., 28 I.

April 10 I., 11 I., 17, 18 I., 19, 20, 21, 22 I., 23, 24 I.

May 15, 16 I., 17, 18, 19 I., 22 I., 23 I.

June 5.

July 12 I., 14, 15, 19.

August 25 I., 26 I., 27 I.

November 6 I., 7 I., 10 I., 11 I., 12 I.

1881—

January 16 I.

August 15 I.

December 24 I.

www.ingramcontent.com/pod-product-compliance
Lightning Source LLC
Chambersburg PA
CBHW021959190326
41519CB00010B/1321